Françoise Combes and James Lequeux

# The Milky Way
Structure, Dynamics, Formation and Evolution

**CURRENT NATURAL SCIENCES**

EDP Sciences/CNRS ÉDITIONS

Cover illustration: Messier 109 (also called NGC 3992), a nearby galaxy of similar morphology to the Milky Way, i.e. a clone of the Milky Way. This barred spiral galaxy gives the right impression of how might look our Galaxy, if seen face-on. © *NOAO/AURA/NSF*.

Printed in France

© **2016, EDP Sciences**, 17 avenue du Hoggar, BP 112, Parc d'activités de Courtabœuf, 91944 Les Ulis Cedex A

et

**CNRS Éditions**, 15, rue Malebranche, 75005 Paris.

This work is subject to copyright. All rights are reserved, whether the whole or part of the material is concerned, specifically the rights of translation, reprinting, re-use of illustrations, recitation, broad-casting, reproduction on microfilms or in other ways, and storage in data bank. Duplication of this publication or parts thereof is only permitted under the provisions of the French Copyright law of March 11, 1957. Violations fall under the prosecution act of the French Copyright law.

ISBN EDP Sciences: 978-2-7598-1915-7
ISBN CNRS Éditions: 978-2-271-09168-0

# Preface

What is this bright band across the sky? Although Democritus was already thinking, in the 5th century BC, that the Milky Way was "made of tiny heavenly bodies grouped so closely that they seem to us to be one" (Achilles Tatius, quoted by Jean Salem, "Democritus", our English translation), it was not until Galileo and his telescope that this bold idea was confirmed. Subsequently, the major obstacle to interpreting the observations, even of excellent quality, in order to establish the size of our Galaxy and the Sun's position within it, was the poor determination of distances. It was only in the 1930s that a correct representation of the Galaxy was obtained, showing that the Milky Way was a galaxy among others, with a radius of 15 kpc (45,000 light years) for its stellar component, of about 20 kpc for its gas component, and that the Sun was far from being at its center.

During the last two decades, new means of observation and new computing facilities have opened new horizons: the advent of space astrometry with the Hipparcos satellite of the European Space Agency (ESA) and its high precision astrometric measurements for more than 100,000 bright stars and very precise distances to 30,000 stars has led to a thorough knowledge of the solar neighborhood and to revised cosmic distance scales; systematic photometric observations over large areas of the sky such as the Sloan Digital Sky Survey (SDSS) have led to the discovery of new stellar streams in the halo; high-resolution spectroscopic observations with large telescopes have led to a much better understanding of the chemical evolution of the Galaxy; observation of millimeter and sub-millimeter waves has led to the discovery of many new molecules in the interstellar medium; and finally increasingly powerful computers have allowed increasingly detailed simulations of the formation and evolution of galaxies.

The coming decade is once again full of promise with the operation of satellites, telescopes and radio telescopes even more sensitive and / or more accurate than their predecessors.

In the optical domain, the ESA Gaia satellite, successor to Hipparcos and second astrometric satellite, was launched in December 2013. It will allow a fantastic step forward in the knowledge of all the stellar components of our Milky Way, with the identification and systematic measurement of one billion objects brighter than magnitude 20, with astrometric precision still 50 to 100 times higher than that of Hipparcos and parallel observation of their physical characteristics. Also in the optical domain, planned for the early 2020s, the E-ELT (European Extremely Large Telescope) will observe, in very great detail, very faint objects in our Galaxy and far beyond.

In the infrared, submillimeter and millimeter domains, essential information is obtained about the formation of stars. After the spectacular results of the Herschel European satellite, the mission of which completed in June 2013, ALMA (Atacama Large Millimeter / submillimeter Array), the global network observing in the millimeter wavelengths, has become fully operational. Later, the JWST (James Webb Space telescope), observing in the near-infrared, is due to be launched in 2018 with the largest telescope ever put into orbit, 6.5 m in diameter.

Finally, in the radio domain, extremely powerful for the study of the interstellar medium and in particular the gas, the first light from the SKA (Square Kilometer Array) project is expected in the 2020s.

A new golden age for astronomy, especially for the study of our Milky Way and Local Group galaxies, the next decade promises to be full of surprises and discoveries, and this book is precisely issued at the right time to focus on our present knowledge before these new steps.

With the precision achievable by space astrometry, this ancient specialty is now a vital tool for astrophysics (in the sense of the physical analysis of the sources observed). It brings cosmic distance scales both for the stellar and gaseous components, and the motions of stars in the solar neighborhood. Soon, thanks to Gaia, these data will be available all across the Milky Way and nearby galaxies. These observations provide clues to the structure of the Galaxy and of its various components, but also to the kinematics and dynamics of these, leading, for example, to a complete description of the orbits of the stars in the Galaxy. Various correlations may now be studied between the orbital characteristics (eccentricity, mean velocity, velocity dispersion) of carefully selected groups of stars and the abundances of chemical elements in their atmospheres. Only the combined study of these parameters allows interconnection of the various traces left by the successive steps of the formation and evolution of our Galaxy.

Astronomers are making progress in the understanding of our Milky Way by assembling the various parts of this puzzle, by comparing these to the characteristics of external galaxies, and by confronting all these observations to increasingly detailed numerical simulations. Conversely, the Milky Way is, of course, the galaxy studied in the highest detail (very accurate distances and motions for many different types of stars, detailed abundances of chemical elements in their atmosphere, detailed description of star forming regions, determination of star orbits very close to the central black hole, only to quote a few), and this provides an essential lighting in the interpretation of much more global observations available for other galaxies.

The book of Francoise Combes and James Lequeux takes us step by step through this rapidly evolving field, with a fascinating description of the present state of our knowledge. The two authors, internationally recognized specialists of the dynamics of galaxies and of the interstellar medium, both have a very broad culture in astronomy and perfect clarity of presentation. They are already the authors of many books on astronomy for the specialist as well as for the general public. This book will certainly become a reference in the field. It is a remarkable introduction to the description of this set of stars, gas and dust in which we live: Françoise Combes and James Lequeux introduce here these complex topics in a form that is concise but very educational, simple but thorough and rigorous. Student, specialist or simply curious, this book will encourage the reader to further deepen their knowledge and push some, I am sure, to embark on the adventure of research and of the interpretation of the mass of data expected from the future instruments of the 21st century.

Catherine TURON<br>
Astronomer Emeritus at the Paris Observatory

The authors thank the experts appointed by the CNRS and Dr Florian Gallier and Dr Jacques Uziel from Cergy Pontoise University, Dr Isabelle Billault from Paris-Sud University and Prof. Alberto Marra from Montpellier University for their careful proofreading.

# Contents

| | |
|---|---|
| Preface | iii |
| Physical and astronomical constants | ix |

**1 Introduction**   1
- 1.1 Shape and dimensions of the Milky Way . . . . . . . . . . . . . . . . . 1
- 1.2 Rotation and spiral structure . . . . . . . . . . . . . . . . . . . . . . . . 6
- 1.3 The Milky Way at all wavelengths . . . . . . . . . . . . . . . . . . . . . 10
- 1.4 The role of the HIPPARCOS satellite . . . . . . . . . . . . . . . . . . . 12

**2 The solar neighborhood**   17
- 2.1 The fundamental parameters of stars and the Hertzprung-Russell diagram . . . . . . . . . . . . . . . . . . . . . . . . . . . . . . . . . . . . 17
- 2.2 The local stellar disk . . . . . . . . . . . . . . . . . . . . . . . . . . . . . 21
- 2.3 Kinematics and dynamics of the stars of the local disk . . . . . . . . . . 25
- 2.4 High-velocity stars . . . . . . . . . . . . . . . . . . . . . . . . . . . . . . 30
- 2.5 The interstellar matter near the Sun . . . . . . . . . . . . . . . . . . . . 31

**3 Structure and components of the Milky Way**   37
- 3.1 Dimensions and rotation of the Galaxy . . . . . . . . . . . . . . . . . . . 37
- 3.2 Stellar populations in the Galaxy . . . . . . . . . . . . . . . . . . . . . . 44
  - 3.2.1 The stellar halo . . . . . . . . . . . . . . . . . . . . . . . . . . . 46
  - 3.2.2 The bulge . . . . . . . . . . . . . . . . . . . . . . . . . . . . . . 47
  - 3.2.3 The thick disk . . . . . . . . . . . . . . . . . . . . . . . . . . . 48
  - 3.2.4 The thin disk . . . . . . . . . . . . . . . . . . . . . . . . . . . . 50
- 3.3 The interstellar medium in the Galaxy . . . . . . . . . . . . . . . . . . . 51
  - 3.3.1 The atomic "neutral" medium . . . . . . . . . . . . . . . . . . . 51
  - 3.3.2 The molecular medium and the interstellar dust . . . . . . . . 55
  - 3.3.3 The ionized medium . . . . . . . . . . . . . . . . . . . . . . . . 60
  - 3.3.4 Supernova remnants, bubbles and hot gas . . . . . . . . . . . . 63
- 3.4 Radiation fields, magnetic field, cosmic particles and radio radiation . . . . . . . . . . . . . . . . . . . . . . . . . . . . . . . . . . . . 64
- 3.5 The spiral structure of the Galaxy . . . . . . . . . . . . . . . . . . . . . 72
- 3.6 Dark matter in the Galaxy . . . . . . . . . . . . . . . . . . . . . . . . . 75
  - 3.6.1 The contribution of baryons . . . . . . . . . . . . . . . . . . . . 77
  - 3.6.2 A gas contribution? . . . . . . . . . . . . . . . . . . . . . . . . 79
  - 3.6.3 Distribution of dark matter in the Galaxy . . . . . . . . . . . . 80
  - 3.6.4 An alternative possibility: modified gravity . . . . . . . . . . . 82

| 4 | The galactic center | 85 |
|---|---|---|
| | 4.1 Bar and bulge | 85 |
| | 4.2 The interstellar matter at the galactic center | 87 |
| | 4.3 The black hole | 90 |
| |     4.3.1 The close environment of the black hole | 90 |
| |     4.3.2 Flares near the black hole | 93 |
| |     4.3.3 The black hole itself | 94 |
| |     4.3.4 Gas in-falling onto the black hole | 98 |
| | 4.4 Conclusion | 101 |
| **5** | **Galactic dynamics** | **103** |
| | 5.1 Dynamics of the barred spiral structure | 103 |
| | 5.2 Cycle of the bar evolution, migrations, multiple waves | 109 |
| |     5.2.1 Destruction and re-formation of bars | 110 |
| |     5.2.2 Migrations | 112 |
| |     5.2.3 Secondary bar, multiple waves | 113 |
| **6** | **The chemical evolution of the Galaxy** | **121** |
| | 6.1 The formation of the Galaxy | 122 |
| | 6.2 The production of elements in stars | 123 |
| | 6.3 Modeling the chemical evolution | 127 |
| | 6.4 The chemical evolution of the halo and the bulge | 131 |
| | 6.5 The chemical evolution of disks | 134 |
| **7** | **Formation and evolution of the Galaxy** | **139** |
| | 7.1 The thin and thick disks | 139 |
| | 7.2 The formation of the bulge | 142 |
| | 7.3 The formation of the halo : cosmological or not? | 144 |
| **8** | **The Galaxy among its companions** | **147** |
| | 8.1 A spiral among the spirals – the Hubble classification of the Galaxy | 147 |
| | 8.2 The satellites : the Magellanic Clouds and dwarf elliptical galaxies | 149 |
| | 8.3 Capture of the Sagittarius dwarf, and many others: the tidal streams | 152 |
| | 8.4 Galactic wind, high velocity clouds, cosmic accretion | 154 |
| | 8.5 APPENDIX | 157 |
| | 8.6 List of the principal Milky Way satellites, sorted by increasing distance | 157 |
| **9** | **The future** | **159** |

**Appendix 1. Stellar parameters**     **165**

**Appendix 2. A few basic notions concerning the observations of the interstellar medium**     **167**

**Glossary**     **169**

**Bibliography**     **177**

**Index**     **181**

# Physical and astronomical constants

| | |
|---|---|
| Astronomical unit | AU = $1.496 \times 10^{11}$ m |
| Light year | ly = $9.46 \times 10^{15}$ m |
| Parsec | pc = $3.086 \times 10^{16}$ m = 3.262 ly |
| Solar mass | $M_\odot$ = $1.989 \times 10^{30}$ kg |
| Solar luminosity | $L_\odot$ = $3.845 \times 10^{26}$ W |
| Tropical year | year = 365.242 days = $3.156 \times 10^{7}$ s |
| Light velocity | c = $2.997\,924\,58 \times 10^{8}$ m s$^{-1}$ |
| Gravitation constant | G = $6.673 \times 10^{-11}$ N m$^{2}$ kg$^{-2}$ |
| | = $6.673 \times 10^{-8}$ dyne cm$^{2}$ g$^{-2}$ |
| Planck's constant | h = $6.626 \times 10^{-34}$ W s$^{-1}$ |
| Boltzmann's constant | k = $1.381 \times 10^{-23}$ W K$^{-1}$ |
| Stefan-Boltzmann's constant | s = $5.671 \times 10^{-8}$ W m$^{-2}$ K$^{-4}$ |
| Mass of electron | $m_e$ = $9.109 \times 10^{-31}$ kg |
| Mass of proton | $m_p$ = $1.673 \times 10^{-27}$ kg |
| Rydberg energy | ryd = $2.180 \times 10^{-18}$ J = 13.606 eV |
| Wavelength associated to 1 rydberg | 91.176 nm |
| Mass energy of electron | 0.511 MeV = $8.187 \times 10^{-14}$ J |
| Mass energy of proton | 938 MeV = $1.503 \times 10^{-10}$ J |

# Units and conversion

**Length**
meter (I.S. unit)  m = 100 cm
angström  A = $10^{-8}$ cm = $10^{-10}$ m

**Mass**
kilogramme (I.S. unit)  kg = $10^{3}$ g

**Energy**
joule (I.S. unit)  J = $10^{7}$ erg

**Power**
watt (I.S. unit)  W = $10^{7}$ erg s$^{-1}$

**Flux density**
jansky (I.S. sub-unit)  Jy = $10^{-26}$ W m$^{-2}$ Hz$^{-1}$ = $10^{-23}$ erg s$^{-1}$ cm$^{-2}$ Hz$^{-1}$

**Force**
newton (I.S. unit)  N = $10^{5}$ dyne

**Pressure**
pascal (I.S. unit)  Pa = N m$^{-2}$ = 10 dyne cm$^{-2}$ = $10^{-5}$ bar

**Magnetic field or induction**
tesla (I.S. unit)  T = $10^{4}$ G (gauss)

# Chapter 1
# Introduction

The luminous band of the Milky Way (our galaxy, also named *the Galaxy*), which crosses the sky as a scarf, has been the object of many myths since prehistoric times. It was considered by the Greeks as due to milk escaped from Hera's breast as she refused to feed Heracles, discovering that he was not her son: hence the name of the Milky way, which is still in use. During the Middle ages, it was Saint-Jacques's path, supposed to orient the pilgrims on their way to Saint-Jacques of Compostelle. Claude Ptolémée (ca.90 – ca.168) produced a detailed description of the Milky Way, which remained unsurpassed for a long time. However, the true nature of the Milky Way was only revealed in 1610 by Galileo (1564-1642), whose astronomical telescope resolved for the first time its diffuse light into many individual stars: he wrote "The Milky Way is just a cluster of innumerous stars". Actually, all the stars and planets that we see in the sky belong to the Milky Way, and the only two objects which do not belong to it are the two Magellanic Clouds in the southern sky and the Andromeda galaxy in the northern sky.

## 1.1 Shape and dimensions of the Milky Way

One had to wait for a century and a half after Galileo to have the first ideas on the shape and the size of the Milky Way. Thomas Wright (1711-1786), in his 1750 book entitled *An original Theory or new Hypothesis of the Universe*, described the Milky Way as a flat stellar system inside which we are located, a system that would be a part of a gigantic spherical shell. However, this was more inspired by a medieval-type cosmogony than by a real scientific reflection. Others, like Emanuel Swedenborg (1688-1772), Immanuel Kant (1727-1804) and Johann Heinrich Lambert (1728-1777), limited themselves to similar considerations. However, they all considered that the stars of the Milky Way should rotate around some unknown center to ensure the stability of the system. But it was William Herschel (1738-1822) who performed the first serious scientific studies of our Galaxy.

Herschel knew that some stars are not really fixed in the sky, but possess a *proper motion* (lateral displacement). Already, Edmond Halley (1656-1742) had suspected that Aldébaran, Sirius and Arcturus could have a proper motion, and Jacques Cassini (1677-1756) had clearly seen in 1738 the proper motion of Arcturus. In 1783, Herschel, who himself had made new observations of stellar positions, noticed that the dozen proper motions, that were known, corresponded to displacements towards a privileged direction. He concluded that it was in fact the Sun that moved in the opposite direction, the *apex*, in the Hercules constellation. This was the beginning of the kinematical studies of stars. However, the velocity of the solar motion was then unknown, for the lack of distance estimates (it is of the order of 20 km s$^{-1}$, see section 2.3).

Herschel was also the first to attempt to obtain a better geometrical image of the Milky Way, from star counts in various directions. For this, he assumed that all stars have the same intrinsic flux, and thus that their apparent flux decreases as the inverse square of their distance. This allowed him to estimate roughly their distance, at least as a relative value. He also assumed that the number of stars per unit volume was the same everywhere. For him, the faintest observed stars were lying at the limit of the system. He obtained in this way in 1784-85 a 3-D geometrical description of the Milky Way, and represented a cut perpendicular to the Galactic plane of symmetry as shown in Figure 1.1. He claimed that the Milky Way had a size of 800 times the mean distance between the stars in the Galactic plane, and only 150 times in the perpendicular direction. The real dimensions were unknown because no distance of any star had been determined, apart from that of the Sun. What was the ratio of the apparent flux of the Sun and of a bright star like Sirius, and were these stars comparable? The beginnings of an answer to these questions came only during the first half of the 19$^{\text{th}}$ century.

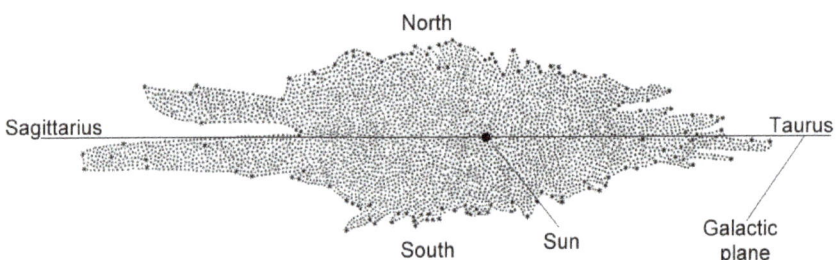

FIG. 1.1 – A cut of the Milky Way perpendicular to its plane of symmetry, as drawn by Herschel. To the left, the lack of stars corresponds to the dark band that splits the Milky Way in the direction of Sagittarius, due to extinction by interstellar dust, something that Herschel could not know. From Herschel, W. (1785) *Philosophical Transactions* 75, 213-266.

However, Herschel had legitimate doubts about the hypotheses he had to make in his work. He realized that stars should exist, fainter than those he could see in his telescopes, and that makes it impossible to determine the real limits of the Milky Way. In his late papers, following on from 1817-18, he admitted that "the Milky Way is unfathomable".

This record of failure slowed the further works, until the Russian astronomer Otto Struve (1819-1905) resumed them on new bases. He acknowledged in 1847 that the density of stars in the Milky Way was far from uniform, contrary to Herschel's hypothesis: it decreases progressively with increasing distance from the Galactic plane. Now, some stellar distances were available, allowing the dimensions of the Milky Way to be obtained: Struve claimed that they were at least $8.17 \times 10^8$ astronomical units, i.e. $1.2 \times 10^{17}$ km, or 13 000 light years, or 4 000 parsecs[1]. Finally, Struve suspected the possibility of interstellar extinction which would reduce the light from a star faster than the inverse square of its distance.

The next important step in the description of the Galaxy came from the Dutch astronomer Jacobus Cornelius Kapteyn (1851-1922), who made his laboratory in Groningen the main center of galactic studies worldwide. He had, at his disposal, photographs of the sky, deep and relatively complete stellar catalogues, and a number of determinations of proper motions and of radial velocities (the velocities of stars along the line of sight, as measured from the displacement of spectral lines using the Doppler-Fizeau effect). In 1906, he launched a large international project for the study of the distribution of stars in the Galaxy, consisting in systematically measuring the magnitudes, the proper motions and the radial velocities of stars in 206 zones of the sky, the *selected areas*. In the meantime, before the completion of this project which implied the cooperation of more than 40 different observatories, Kapteyn started his own study of the distribution of stars in the Milky Way. Now, he could account for the different intrinsic luminosity of the stars, which he described by a *luminosity function*. But this yielded a new difficulty: the distribution of the apparent magnitudes of stars resulted from the combination of their different luminosities and of their different distances. Kapteyn succeeded in solving this problem in a very ingenious way. He illustrated his results in the schematic form of Figure 1.2, which corresponds to his final model of 1922. For him, the Galaxy was a flattened ellipsoidal system, in which the Sun occupied a slightly eccentric position. This model was more schematic than that of Herschel, but represented considerable progress by showing how the density of stars decreases to the exterior of the Galaxy, and by the introduction of a distance scale.

---

[1] The astronomical unit (a.u.) is the half-major axis of the Earth's orbit, $1.496 \times 10^{11}$ m. The parsec is the distance from which this half-major axis is seen under an angle of 1 arc second: 1 pc = 206 285 a.u. = $3.086 \times 10^{16}$ m = 3.26 light-years.

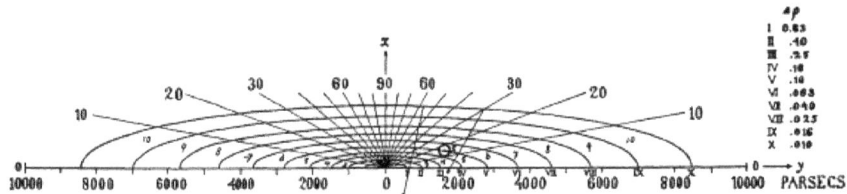

FIG. 1.2 – The Galaxy according to Kapteyn in 1922. It was schematized by a series of concentric ellipsoids, whose density decreased to the exterior according to the scale at the right of the figure. The circle represented the position of the Sun. From Kapteyn, J.C. (1922) *Astrophysical Journal* 55, 302-328, with the permission of the American Astronomical Society.

However, Kapteyn's model was wrong, because, similar to all his predecessors, he did not take into account interstellar extinction. Curiously, he had supposed the existence of extinction in his first works, but he rejected it later. In 1904, Johannes Franz Hartmann (1865-1936), at the Potsdam astrophysical observatory, had noticed in the spectrum of the star δ Orionis very narrow absorption lines that he attributed to calcium ions located in intervening gas clouds. In 1912, the American astronomer Vesto Slipher (1875-1969) discovered the interstellar dust grains illuminated by the light of the Pleiades stars, and suggested that this dust could well absorb the light of background stars. Finally, photographs by Edward E. Barnard (1857-1923) and Max Wolf (1863-1932) had shown the existence of regions of the Milky Way apparently devoid of stars, and this was attributed at the end of the 1910s to dark dust clouds. One then started to interpret the dark band that seems to split the Milky Way not by the absence of stars, but by extinction by dust.

This allowed the Swiss-American astronomer Robert J. Trumpler (1886-1956) to give, in 1930, a definitive description of the Galaxy. Trumpler noticed first that the angular diameter of the distant open clusters[2], which are close to the galactic plane, looked abnormally large if they were at the distance derived from their luminosity without any correction. But if an interstellar extinction exists, their distance is in fact smaller and everything returns to normal. Trumpler derived from this a numerical value for extinction by unit distance in the Galactic plane.

Next he examined the distribution of globular clusters of stars, the majority of which are far from the Galactic plane: their light is not affected much by interstellar extinction, which is clearly concentrated along the plane. Harlow Shapley (1885-1972) had shown previously that most of these clusters lie in one half of the sky and formed a spherical system whose center

---

[2] See the end of this chapter for illustrated definitions of the different objects encountered in the Milky Way.

was far from the Sun, in the direction of the Sagittarius constellation. He had estimated their distance thanks to the variable stars they contain (the RR Lyrae) and concluded that if they really belonged to the Milky Way, the center of their system should also be the center of the Galaxy, at a distance of about 20 000 parsecs. Trumpler, and then Joel Stebbins (1878-1966) and Albert Whitford (1905-2002) in 1936, revised this distance to 8 000 pc, a value confirmed by recent estimates. From all these studies resulted a model of the Galaxy represented in Figure 1.3, which is still completely valid today.

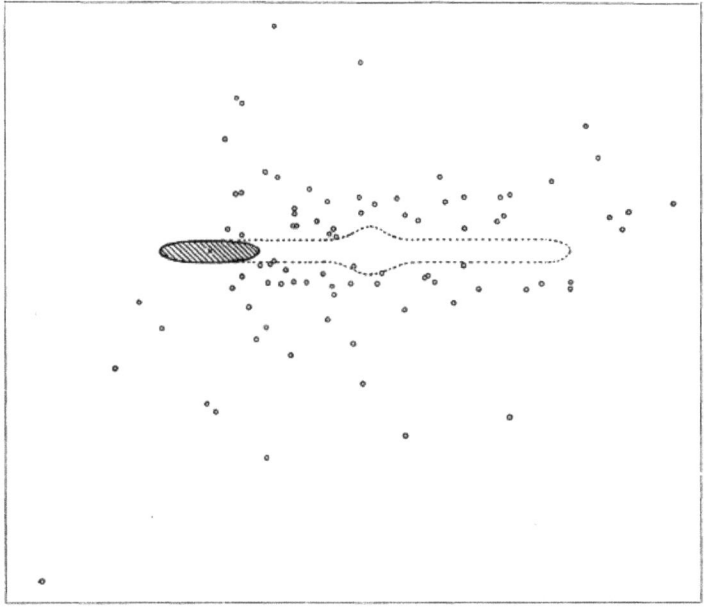

FIG. 1.3 – A cut of the Galaxy, according to Shapley, Trumpler, Stebbins and Whitford. The dotted contour encompasses most of the stars and interstellar matter. The hatched ellipse is Kapteyn's Galaxy, limited by interstellar extinction, with the Sun almost at its center. The small circles symbolize the globular clusters. From Trumpler, R.J. (1941) *Publications of the Astronomical Society of the Pacific* 53, 155-165, with permission of the Editor.

The astronomers at the time noticed that the Galaxy is rather similar to the Andromeda nebula and many similar objects. They became fully aware that the Milky Way is a galaxy similar to many others, and also that the Sun is far from its center, in a remote region.

## 1.2 Rotation and spiral structure

Let us say now a few words about the motions in the Galaxy. After enough radial velocities of globular clusters and external galaxies had been measured in the 1920s, it became clear that all the stars near the Sun move with an enormous velocity, about 300 km/s, with respect to the average of these objects: this was the discovery of the rotation of the Galaxy, which keeps its different parts, in particular the solar neighborhood, in equilibrium between the gravitational attraction of the central regions and the centrifugal force. The Swedish astronomer Bertil Lindblad (1895-1965) and his Dutch colleague Jan Oort (1900-1992) then showed that the Galactic disk does not rotate as a solid body, but that the regions closer to the center rotate faster than the external regions: this is the *differential rotation*. They could understand in this way a phenomenon discovered previously by Kapteyn. Kapteyn had observed that the stars near the Sun move along two opposed currents perpendicular to the direction of Sagittarius, which is that of the Galactic center. These two currents are a consequence of the differential rotation.

Thanks to the galactic rotation, it became possible to determine its mass. In this context, a major event for galactic astronomy, and for astronomy in general, occurred in 1951: the discovery of the radio emission of atomic interstellar hydrogen at the wavelength of 21 cm, the *21-cm line*. Predicted by the Dutch physicist Hendrick van de Hulst (1918-2000) and discovered in the USA by Harold I. Ewen (born 1922) and Edward M. Purcell (1912-1997), this line allowed, for the first time, observation of the whole Galaxy, because there is no interstellar extinction of radio waves. The radial velocity of the emitting regions can be obtained from the Doppler-Fizeau line shift. This makes it possible to determine the rotation velocity in the Milky Way as a function of the distance to the Galactic center (the rotation curve) and to draw the first complete map of the interstellar gas in the Galaxy (Fig. 1.4), which is dominated by hydrogen. Spiral arms can be seen over a large extent, while only the nearest ones could be suspected by optical observations: this confirmed the similarity of our Galaxy with external spiral galaxies.

In 1970, the discovery of radio lines of the interstellar CO (carbon monoxide) molecule opened new horizons for the knowledge of the Galaxy. This molecule is a good tracer of molecular gas, while it is difficult to observe the hydrogen molecule $H_2$. Much effort has been devoted to observe the CO lines at 2.6 and 1.3 millimeter wavelengths. Figure 1.5 is a comparison between an image of the inner half of the Milky Way and a map in the 2.6-mm CO line: there is a perfect correspondence between the absorption features due to interstellar dust and the molecular gas. Like for the 21-cm line, it is possible with the CO lines to obtain the distance of the emitting regions and thus to map the molecular gas. Its total mass is larger than that of the atomic gas.

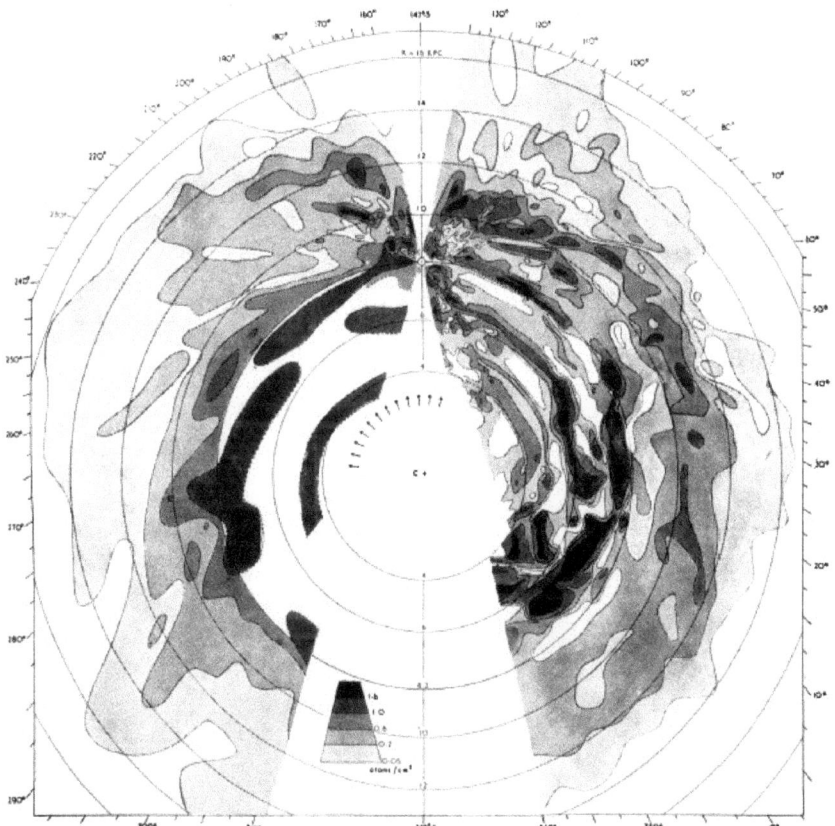

FIG. 1.4 – The first complete map of the Galaxy in the 21-cm line of atomic interstellar hydrogen. C is the Galactic center. The Sun is at 8 kpc above. The surface density of hydrogen is given by the gray levels. The spiral structure is visible, but the details are uncertain because the distances are obtained from the radial velocities assuming pure rotation, although there are important local velocity deviations. The system of galactic longitudes used in this map is obsolete. From Oort, J.H., Kerr F.T. & Westerhout, G. (1958), *Monthly Notices of the Royal Astronomical Society*, 118, 379-389, Wiley, with permission of the Editor.

Radioastronomy – the study of the Universe in radio waves – is also useful for observing gaseous nebulae[3]. They emit not only a continuum and emission lines in the visible, but also in radio. The wavelength shift of these lines gives the radial velocity. The radio observation allows information to

---

[3] Also called HII regions, because they mainly contain ionized hydrogen. Astronomers use to designate the various degrees of ionization by the roman figures I for neutral, II for singly ionized, III for doubly ionized, etc.

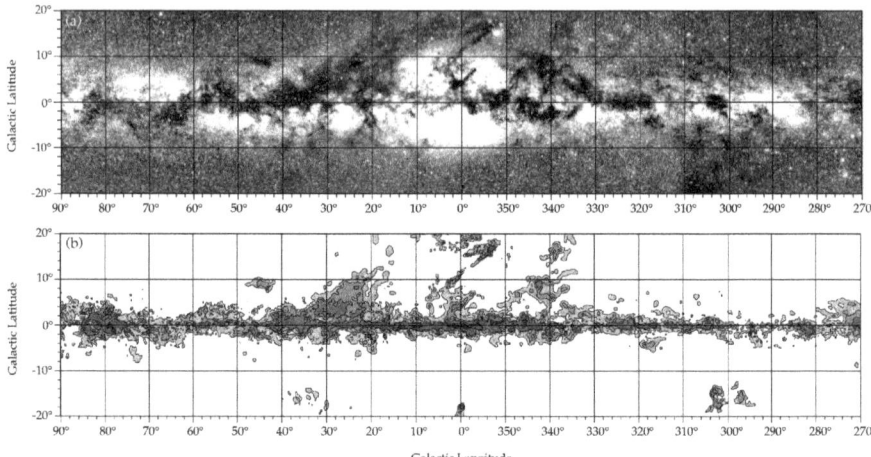

FIG. 1.5 – Comparison of extinction by interstellar dust and the distribution of the interstellar CO molecule. Top, a photographic mosaic of the half of the Milky Way centered on the direction of the Galactic center, which is at the origin of the coordinates. Bottom, a map in the 2.6-mm line of CO. The correspondence is generally excellent, showing that the molecular gas and the dust are well mixed. However, some dust does not correspond to molecular gas, and is associated with atomic or ionized gas. From Dame, T.M., Hartmann, D. & Thaddeus, P. (2001) *Astrophysical Journal* 547, 792-813, with permission of the American Astronomical Society.

be obtained on distant nebulae that are not visible optically, and derivation of their distance from their radial velocity. The observation of external spiral galaxies shows that gaseous nebulae are excellent tracers of spiral arms. In our Galaxy, Yvon and Yvonne Georgelin obtained in 1976, from visible and radio observations of gaseous nebulae, a map of the spiral arms of the Milky Way: an updated version is reproduced in Figure 1.6. Figure 1.7 is a photograph of an external galaxy that is generally considered as a twin of our own Galaxy.

What is the origin of the spiral structure? Since its discovery by Lord Rosse (1800-1867) in the middle of the 19th century, the question has been continuously raised. The discovery of the differential rotation has made any explanation even more problematic, because the deformation caused to the Galaxy destroys any feature in a time that is short with respect to the age of the Universe: as a consequence, only a small fraction of the galaxies should be spiral if the spiral arms are a material structures driven by the rotation. It became progressively clear that to survive, the arms cannot follow the rotation. A satisfactory solution to the problem of spiral arms was finally given in 1964 by the Sino-American astronomers Chia-Chiao Lin and Frank

Introduction

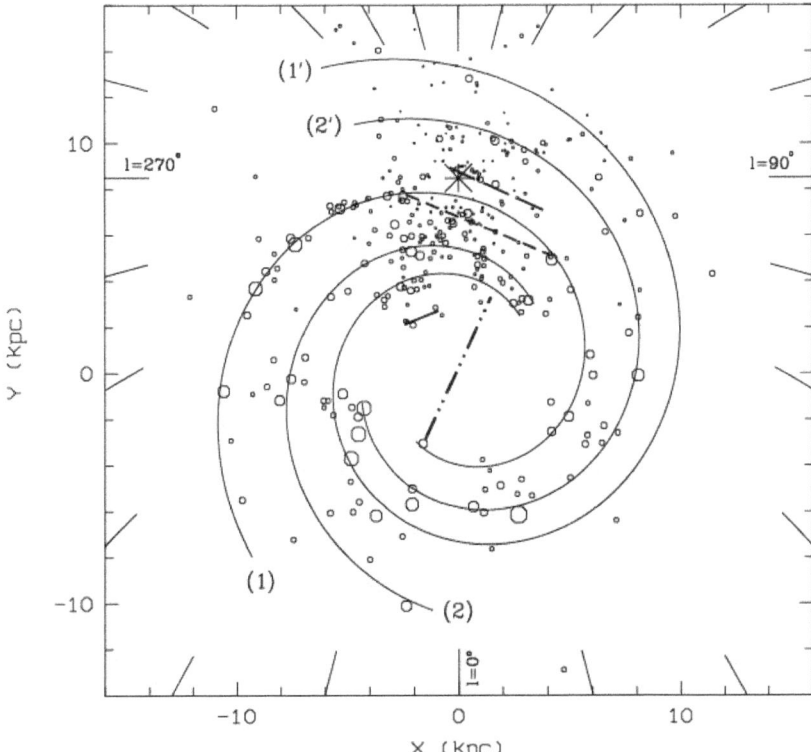

FIG. 1.6 – A map of the Galactic spiral arms obtained from observations of gaseous nebulae. The position of the Sun, here supposed to be at 8.5 kpc from the Galactic center, is represented by a star symbol. The circles represent nebulae with known distances; their size is linked to the far-ultraviolet flux of their ionizing stars. The best fit corresponds to a 4-arm logarithmic spiral. The central bar of the Galaxy is schematized by a dot-dot-dashed line. The local arm is drawn as the long-dashed line, and a foreseen deviation of the inner arm (Sagittarius-Carina) by a short-dashed line. Compare to Figure 1.4, for which the spiral arm pattern is less reliable. See also further Figure 3.4. From Russeil, D. (2003) *Astronomy & Astrophysics* 397, 133-146, with permission of ESO.

Shu : they showed that the arms are temporary compression regions of the material of the galactic disk, i.e. density waves rather similar to sound waves. When the interstellar gas enters such a density wave, its compression favors the formation of molecules and triggers the gravitational collapse of a fraction of the interstellar "clouds". As a consequence, the spiral arms are rich in compressed molecular gas that form stars by collapse. The most massive stars are very hot and produce a large amount of ultraviolet radiation, which ionize the surrounding gas, forming gaseous nebulae. All this is confirmed by observation. In particular, deviations from pure rotation are observed for the gas penetrating in a density wave. How these density waves are formed will be examined in Chapter 5.

FIG. 1.7 – The galaxy NGC 6744, a twin of our Galaxy. The gaseous nebulae are red. The spiral arms are numerous and poorly defined. Notice the bar (horizontal in the picture) and the ring in the central regions. © ESO.

## 1.3 The Milky Way at all wavelengths

The Milky Way has presently been observed at all the wavelengths of the electromagnetic spectrum. Figure 1.8 shows its emission at various wavelengths, with an indication of the corresponding emission mechanism. Note that the interstellar medium is transparent for radio and infrared wavelengths, becomes increasingly opaque from the near infrared to the visible, ultraviolet and soft X-rays, and is again transparent for hard X-rays and gamma rays. Note also that interstellar dust absorbs about one half of the energy emitted by stars, and radiates it in the far infrared.

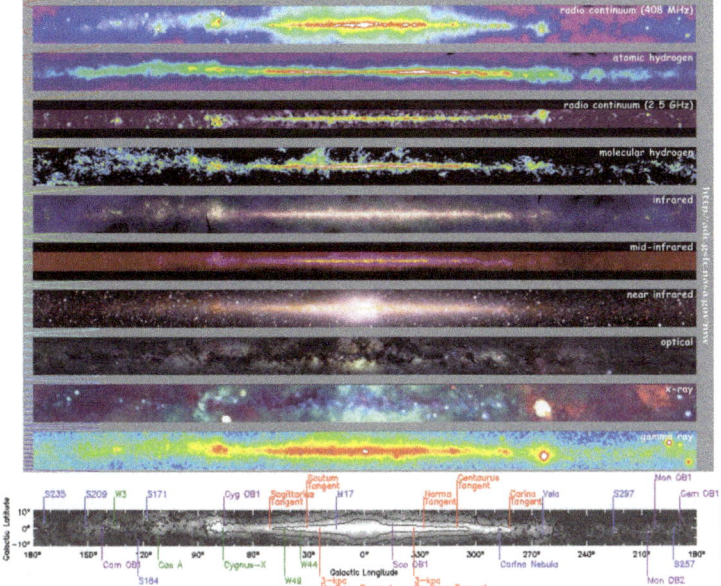

FIG. 1.8 – The multiwavelength Milky Way (© NASA-Goddard Space Flight Center). The images in false colors show a band of ± 10° on either side of the Galactic plane. The Galactic center is at the middle, and the galactic longitude increases from −180° to +180° from left to right. The image at the bottom allows identification of the main features of the Milky Way: arms seen tangentially in red, OB associations in violet, gaseous nebulae in blue and supernova remnants in green. The emission mechanisms are detailed below. For details on the images and their origin, see http://mwmw.gsfc.nasa.gov/mmw_sci.html.
– Radio continuum (408 MHz): synchrotron emission of relativistic electrons (parts of cosmic rays) in the Galactic magnetic field;
– Atomic hydrogen: traced by the 21-cm line emission;
– Radio continuum (2.5 GHz): free-free emission of the ionized interstellar gas;
– Molecular hydrogen: traced by the line emission of CO at 115 GHz;
– Infrared: thermal emission of interstellar dust;
– Mid-infrared: mostly emission of polycyclic aromatic hydrogenated molecules (PAHs);
– Near infrared: emission of stars, unaffected by interstellar extinction;
– Optical: emission of stars, affected by interstellar extinction;
– X-ray: emission of very hot interstellar gas and of some stars, partly affected by absorption by the interstellar gas;
– Gamma ray: emission by the interaction between high-energy "cosmic-ray" protons with interstellar nuclei, and by the braking radiation of cosmic electrons in the field of the nuclei.

## 1.4 The role of the HIPPARCOS satellite

During the last twenty years, much progress has been made on the knowledge of our Galaxy thanks to large stellar surveys, infrared-submillimeter astronomy satellites and the HIPPARCOS astrometry satellite launched in 1989 by the European Space Agency. Many references will be made in this book to stellar surveys and infrared satellites, but as the contribution of HIPPARCOS is more specialized, we will now say a few words about it.

There were previously extensive studies of the proper motion and distance of many stars, but they were superseded by those of HIPPARCOS, which obtained the distance of 120 000 stars with an accuracy better than 1% for those located within 50 parsecs, and better than 10% up to a distance of 250 pc, and even of 400 pc for a fraction of them. The proper motions have been obtained either by the satellite itself, which was operational for more than three years, or by a combination of the HIPPARCOS positions with positions measured earlier from the ground. Moreover, the positions of two and a half million stars have been measured with less accuracy, and their magnitudes have also been obtained. These observations have deeply renewed our knowledge of the vicinity of the Sun, and have supplied a better database for obtaining distances in the Universe. The radial velocities that are necessary to determine the 3-D motion of stars were lacking, but many have been measured in complementary observations.

HIPPARCOS has a successor, GAIA, the second astrometry satellite of the European Space Agency, launched on 19 December 2013. It is presently measuring the characteristics of more than a billion celestial objects up to magnitude 20. The position accuracy reaches 7 millionths of an arc second for the brightest stars (magnitude less than 10), a hundred times better than that of HIPPARCOS. The magnitudes reached are faint enough to cover the whole Galaxy. GAIA also contains a spectrometer able to measure the radial velocity of millions of relatively bright stars. The first catalogue will be released mid-2016, but many years will be necessary to harness the full benefit of this mission, which will certainly renew once again our knowledge of the Milky Way.

### E1.1. Star clusters and nebulae

We give here an illustrated definition of the different types of objects encountered in the Milky Way, apart from the stars themselves.
*Open cluster* (also called *Galactic cluster*): a group of stars born approximately at the same time, located close to the Galactic plane and generally young, because these stars most often are dispersed with time. Some loose, very young clusters, are called *O and B associations* (Fig. E1.1).

FIG. E1.1 – The open cluster NGC 3603. This cluster is the most massive of the Galaxy, and should become a globular cluster in a few hundred million years, after its most massive and luminous stars have disappeared. The image is in the infrared, because the large interstellar extinction in this direction prevents observation of the cluster in the visible range. Image dimensions: 2.1×2.1 pc at the distance of the cluster. © ESO.

*Globular cluster*: a group of stars born approximately simultaneously, whose large total mass has insured stability with respect to dispersion: only the densest and richest open clusters end up as globular clusters after their most massive stars have disappeared. The globular clusters in our Galaxy are generally very old and form an approximately spherical system (Fig. E1.2).

FIG. E1.2 – The globular cluster 47 Tucanae (or NGC 104). Located in the southern sky, this cluster is one of the most spectacular of the Galaxy. Image dimensions: 10.0×10.0 pc at the distance of the cluster. © ESO.

*Gaseous nebula* (also *HII region*): a mass of interstellar gas ionized by hot, massive stars (Fig. E1.3 bottom).

FIG. E1.3 – Bottom, the Trifid Nebula is a gaseous nebula ionized by the ultraviolet radiation of internal stars. Its red color is due to the Hα line of hydrogen, which dominates the visible light. Top, a reflection nebula whose dust is illuminated by a conspicuous star. The scattered light is bluer than that of the illuminating star. Image dimensions: 6.8×11.2 pc at the distance of the nebulae. © ESO.

*Reflection nebula*: a mass of interstellar matter made visible by the scattering by dust of the light of a nearby or internal bright star (Fig. E1.3 top).

*Dark nebula* (also *dark cloud*, or *molecular cloud*): a mass of interstellar mass opaque to light, in which the gas is mainly molecular (Fig. E1.4).

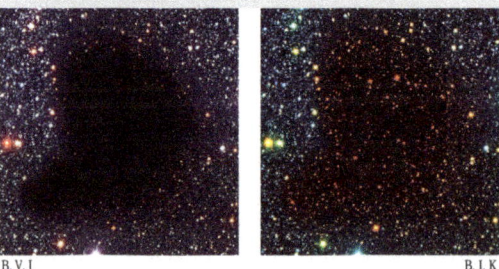

FIG. E1.4 – A small dark cloud, the B68 globule. Left, an image in visible light, in which the light of the background stars is completely extinct by the cloud dust. Right, an infrared image in which the background stars are visible, the extinction by dust being much smaller than in the visible. Image dimensions: 0.22×0.22 pc at the distance of the nebula. © ESO.

*Interstellar cloud*: general term for an identifiable mass of interstellar matter, in which the gas can be mainly atomic (HI cloud) or molecular (*molecular cloud*). The structure of these clouds is generally complex and fragmented, so that the name should not be taken literally (Fig. E1.5).

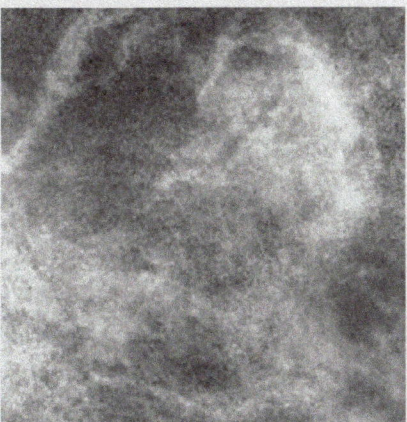

FIG. E1.5 – A map of atomic, neutral hydrogen in a 3°×3° field on the Perseus constellation (approximate dimensions 130×130 pc at the distance of most of the observed gas). Some concentrations could be qualified of clouds, but the map is dominated by sheets and filaments. © Dominion Astrophysical Observatory.

*Planetary nebula*: a mass of gas ejected by a star less massive than about 8 solar masses at the end of its evolution, and partially ionized by the ultraviolet radiation of the very hot remnant of the star (Fig. E1.6).

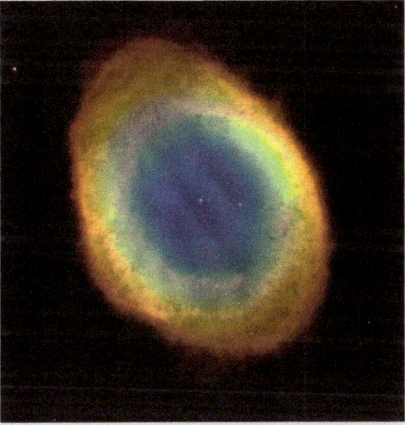

FIG. E1.6 – The planetary nebula M 57. Note the central star that ejected the envelope and ionized it. Image dimensions; 0.36×0.36 pc at the distance of the object. © Hubble Space Telescope Heritage.

*Supernova remnant:* a mass of gas ejected by a massive star during its final explosion; its mass is progressively increased by the interstellar matter encountered during its expansion (Fig. E1.7).

FIG. E1.7 – The supernova remnant Cassiopeia A. The filaments were ejected some 400 years ago by the explosion of a star, which is now invisible. Image dimensions; 4.0×4.0 pc at the distance of the object. © Hubble Space Telescope Heritage.

*Interstellar bubble*: a very large gaseous, spherical expanding shell, which results from the winds emitted by the stars of an open cluster and from the explosion as supernovae of the most massive stars of this cluster (Fig. E1.8).

FIG. E1.8 – The interstellar bubble N70 in the Large Magellanic Cloud, a galaxy satellite of the Milky Way. Image dimensions: 100×100 pc at the distance of the object. The Milky Way contains several such bubbles, but they are more difficult to see due to our unfavorable position. © ESO.

# Chapter 2
# The solar neighborhood

The vicinity of the Sun is obviously the best known part of the Galaxy. We benefit from an almost complete catalogue of stars closer than 25 parsecs, with the exception of the brown dwarfs whose luminosity is extremely low. The distance and proper motion of stars are well determined within some 100 pc, and many radial velocities are known, although in much smaller number than the proper motions. Finally, the local interstellar medium has been explored in detail. However, the solar neighborhood is a relatively quiet place in the Galaxy: it contains no very young, massive star, and the closest gaseous nebula, the Orion Nebula, is at 500 pc from us. Before describing this region, we will provide some notions on stars in general.

## 2.1 The fundamental parameters of stars and the Hertzprung-Russell diagram

The spectral distribution of the radiation of a star is not very different from that of a blackbody; hence the habit of expressing luminosity $L$ of a star, the total energy radiated by it, through an *effective temperature* $T_{eff}$, the temperature of a fictitious blackbody with the same dimensions and luminosity of the star. As a consequence, the following relationship occurs between $L$, $T_{eff}$ and the radius $R$ of the star:

$$L = 4\pi R^2 \sigma T_{eff}^4, \qquad (2.1)$$

$\sigma$ being the Stefan-Boltzmann's constant. It is difficult to obtain $L$, $T_{eff}$ and $R$ by direct observation. An accurate determination of the effective temperature would require the measurement of the full spectral distribution of the stellar radiation from the far-ultraviolet to the mid-infrared, by observations from the ground corrected for atmospheric absorption and observations from space vehicles. One would measure simultaneously the apparent luminosity of the star, i.e. the power received from it per unit area above the Earth's atmosphere. Then the distance would be required to obtain the absolute luminosity $L$.

Fortunately, observation of stars of different types allows determination of a relationship between their color in the visible range and their effective temperature, so that it is sufficient to measure the color to determine an approximate value for $T_{eff}$. The color is defined from measurements through standard colored filters. There are several such photometric systems, but we will only describe here the Johnson system, which is the most classical.

The stellar fluxes are generally expressed in the logarithmic system of magnitudes, which is a quantitative version of the old visual system. If $e(\lambda)$ is the monochromatic flux received above the terrestrial atmosphere at the wavelength $\lambda$, the corresponding magnitude is:

$$m(\lambda) = -2.5 \log e(\lambda)/e_0(\lambda), \qquad (2.2)$$

where the constant $e_0(\lambda)$ defines the zero magnitude. In practice, the measurement is made in a more or less wide spectral band defined by a colored filter, and the above equation has to be integrated on this band. The filters most used for Galactic studies are the blue filter B, centered at 440 nm, and the visual filter V centered at 550 nm, and also the infrared filters J (1.22 μm), H (1.63 μm) and K (2.19 μm).

The color $B-V$ of a star ($B$ and $V$ being the magnitudes in the corresponding filters) and the characteristics of its spectrum used to define its *spectral type*, are related to its effective temperature. Its intrinsic luminosity at some wavelength band, for example in the V filter, is expressed as an *absolute magnitude M*, for example $M_V$ in the V filter. By definition, the absolute magnitude is equal to the apparent magnitude that the star would have outside the atmosphere if it were at a distance of 10 parsecs. Thus, the following relation exists between the apparent magnitude m and the absolute magnitude M:

$$M = m + 5 + 5\log D, \qquad (2.3)$$

$D$ being the distance in parsecs. In particular, this relation applies between the magnitudes $M_V$ and $V$.

Table A1 of Appendix 1 gives the color $B-V$, the effective temperature, the V absolute magnitude $M_V$ and the absolute luminosity $L$ for stars of different types.

To obtain the radius of a star and its absolute luminosity, its distance must be known. There are three different ways for obtaining this distance:

- by measuring its *geometric parallax*. During a year, the star describes an ellipse with respect to very distant objects like galaxies or quasars, a parallax effect due to the orbital motion of the Earth. The semi-major axis of this ellipse, which is the parallax π, is equal to one arc second for a star located at a distance of 1 parsec. The HIPPARCOS European satellite has measured geometric parallaxes of a very large number of stars with a precision better than 0.001": the distance of these stars is known in this way with a precision better 10% to a distance of some 100 pc;

- by measuring the average proper motion of stars which are considered as belonging to a common group (*statistical parallax* method). This proper motion μ, expressed in arc seconds per year, is then directed in a direction opposite to the displacement of the Sun with respect to the average of the neighboring stars (the *apex*): this is what Herschel discovered (see Chapter 1) but he could not give a numerical value to the Sun's velocity. The proper motion becomes smaller and smaller for stars with increasing distances. For stars located at 42 pc in a direction perpendicular to that of the apex, the proper motion is 0.1"/year, an easily measurable quantity. This method does not work for isolated stars, which have too high random motions;

- by comparing the magnitudes of two stars with identical observed characteristics, the distance of the nearer one being known by one of the preceding methods: it is then possible to obtain the distance of the other one, after correction for the effects of interstellar extinction; this is the *photometric parallax* method.

If one plots as a function of the *B-V* color and the absolute magnitude $M_V$ of many stars, one obtains the *Hertzprung-Russell diagram* (HR diagram), from the names of the astronomers who first popularized this kind of diagram. Figure 2.1 shows two versions of the HR diagram for relatively nearby stars. One can see that the representative points of the stars are not distributed at random, but that the majority of them follow a sequence called the *main sequence*, on which the stars spend about nine tenths of their life. Another branch emerges from the main sequence: the luminosity increases with *B-V*, hence with decreasing temperature of the star. Such stars differ from those of the main sequence with the same color by their larger luminosity, hence their larger diameter: these are the red giants, which evolve faster than the main sequence stars. Figure 2.2 allows the identification of the various types of stars in the HR diagram.

The HR color-magnitude diagram *B-V*, $M_V$ can be transformed into a diagram giving $L$ as a function of $T_{\text{eff}}$. The latter diagram is often called the *theoretical HR diagram*. It allows a direct comparison of the observations with the results of the stellar evolution models.

The mass of stars can only be determined accurately by the observation of eclipsing binaries, in which one of the two components passes alternatively in front and behind the other, producing variations of the light received from the system: then, one is ensured that we are approximately located in the plane of the orbit. Measuring the radial velocities of the components using the Doppler-Fizeau effect on their spectral lines, one can apply Kepler's laws to obtain the mass of each component. The comparison of the masses determined in this way for main-sequence stars yields a remarkable result: for main-sequence stars, mass is closely related to luminosity (Fig. 2.3). An important property is a consequence: the lifetime on the main sequence is shorter for more massive stars. This lifetime varies as the ratio of the mass (the energy reserve) to the luminosity (the energy loss). As an example, the lifetime of the Sun on the main sequence is about 10 billions years, while that of a star of mass 10 $M_\odot$ (solar masses), whose luminosity is about 10 000 times higher, is only about 10 millions years.

Table A.2 of Appendix 1 gives the luminosity and effective temperature of stars at the beginning of their sojourn on the main sequence (the zero-age main sequence, see Fig. 2.2), and their lifetime on the main sequence. These quantities are slightly dependent on the chemical composition of the star and are given for a star with a composition close to that of the Sun. Note that the lifetime on the main sequence is of the order of the age of the Universe ($1.4 \times 10^{10}$ years) for a star with mass 0.9 $M_\odot$: as a consequence, all the stars with a smaller mass are still on the main sequence. Those with a larger mass formed some $1.4 \times 10^{10}$ years ago (if any) had time to evolve and to become red giants, then compact objects (white dwarfs, neutron stars or black holes according to their mass). This should be kept in mind when considering the stellar content of the Milky Way.

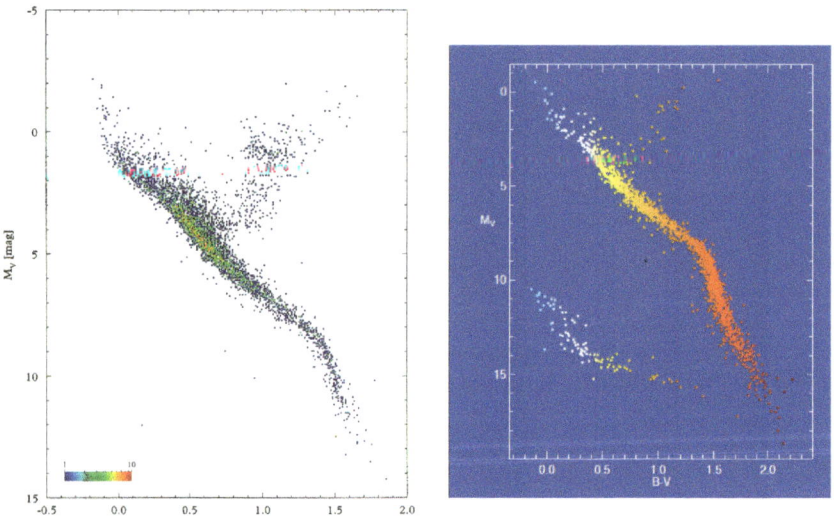

FIG. 2.1 – Left, the Hertzprung-Russell (HR) diagram obtained with the HIPPARCOS satellite for the brightest stars in the sky (© ESA). It contains 4 907 stars whose distances are known within 5%. The colors indicate the representative points containing more than one star. To identify the different types of stars, see Figure 2.2. Right, the HR diagram obtained with HIPPARCOS for 1 924 stars within 25 parsecs from the Sun (the scale for the ordinates is different): *from Jahreiss, H.*, © ESA. This sample is approximately complete, except for the faintest dwarf stars, the white dwarfs and the neutron stars, it gives a better idea of the proportion of the different types of stars: note in particular the paucity of red giants and the large amount of white dwarfs.

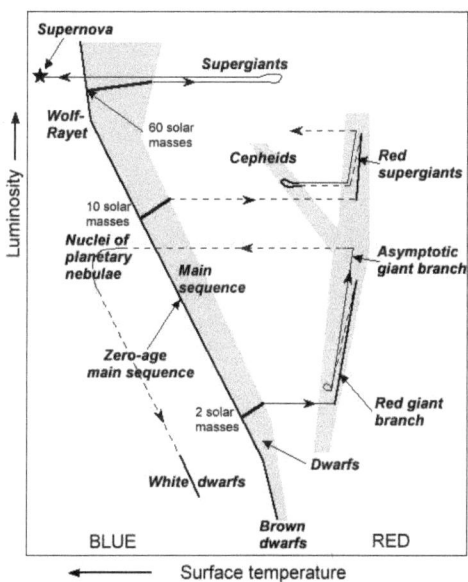

FIG. 2.2 – A Hertzprung-Russell diagram, where one can see in a schematic way where the different types of stars can be found and their evolutionary tracks. Comparing with the HR diagrams of Figure 2.1, relative to the Solar neighborhood, one sees the complete lack near the Sun of very luminous, massive stars: O stars, supergiants, cepheids, Wolf-Rayet stars, etc.

## 2.2 The local stellar disk

The stars in the Galaxy, and in particular in the Solar neighborhood, take part in the differential rotations but also possess random motions, like the particles of a gas. However, their velocity dispersion has no reason to be isotropic as is the case for a collisional gas, and is effectively anisotropic. Moreover, it increases with the age of the stars, which are accelerated by different processes: collision with massive molecular clouds, and above all crossing of spiral arms, which are density waves.

Let us consider first the random motions perpendicular to the Galactic plane. Stars move back and forth through the plane with periodic, but not sinusoidal motions. The restoring force $K_z$ is the gravitational attraction of the Galactic disk, which varies with the distance to its symmetry plane. We will now see how $K_z$ is determined. Once this force is known, it is possible in principle to determine the local mass density $\rho$ using the Poisson equation:

$$\rho = -1/4\pi G \ \partial K_z/\partial z, \qquad (2.4)$$

neglecting for the moment as a first approximation the two other components of the gravitational force, because the ratio of the diameter to the thickness of the disk is very large.

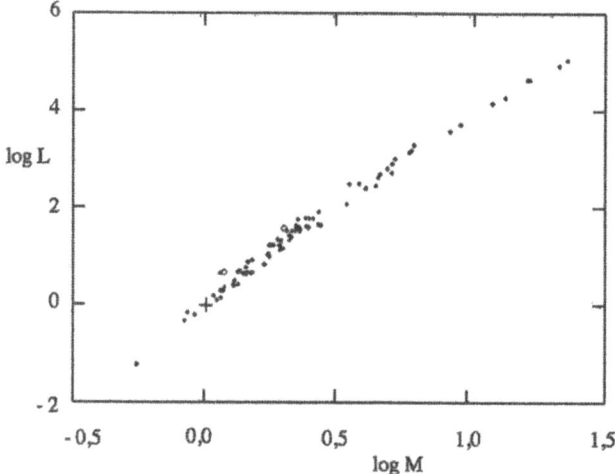

FIG. 2.3 – The mass-luminosity relation for eclipsing binaries for which the spectral lines of both components can be observed. The luminosity and the mass are in solar units, and the place of the Sun is marked by a cross. *From Zahn, J.-P., in Lequeux et al. (2008)*.

A star with vertical velocity $v_0$ in the plane of symmetry reaches the velocity $v(z)$ at the altitude $z$. Applying the theorem of energy conservation, we have:

$$v_0^2 = v_z^2 - 2\int_0^z K_z(z)\mathrm{d}z. \qquad (2.5)$$

In 1932, Oort derived for the first time $K_z(z)$ from statistics on the density of stars as a function of $z$ and from the law of distribution of their velocities. For doing this, it is necessary to assume that the stars form a homogeneous, well mixed ensemble, i.e. that there are as many stars that leave the Galactic plane as stars falling on it, and also that their vertical velocity distribution is independent of $z$ (isothermal "stellar gas"). A further difficulty in the problem is that one should be sure that important masses, e.g. that of a spiral arm, do not perturb the gravitational potential. Finally, one must be sure that the sample of stars under study is complete, or at least that it is possible to correct it for incompleteness, a rather difficult task. The vertical velocity of each star can be obtained directly by measuring the radial velocity of a sample near the Galactic poles, or by measuring the proper motion of stars close to the Galactic plane knowing their distance, which is required in any case. It is also necessary to correct the velocities for the particular velocity of the Sun, and for its distance to the plane, which is fortunately small, about 20 pc to the North of the plane.

Let us assume that all these conditions are met. Let us call $\varphi(z,v)dz\,dv$ the density in the phase space, i.e. the number of stars with a vertical velocity between $v$ and $v+dv$ and a distance to the Galactic plane between $z$ and $z+dz$. The number $n(z)$ of stars per unit volume at the distance $z$ is:

$$n(z) = \int_{-\infty}^{+\infty} \varphi(z,v)dv. \qquad (2.6)$$

If the system is well mixed, the Liouville theorem tells us that the density in the phase space is constant:

$$\varphi(z,v) = \varphi(0,v_0). \qquad (2.7)$$

Let us assume for simplicity that the vertical velocity dispersion is Gaussian with the dispersion $\sigma$ (the problem could of course be solved numerically with any other distribution). We have, at $z = 0$:

$$\varphi(0,v_0) = n(0)[(2\pi)^{1/2}\sigma]^{-1} \exp(-v_0^2/2\sigma^2), \text{ hence} \qquad (2.8)$$

$$\varphi(z,v) = n(0)[(2\pi)^{1/2}\sigma]^{-1} \exp[-v^2/2\sigma^2 + 1/\sigma^2 \int_0^z K_z(z)dz], \text{ then} \qquad (2.9)$$

$$n(z) = n(0) \exp[1/\sigma^2 \int_0^z K_z(z)dz]. \qquad (2.10)$$

Thus, if the velocity dispersion and the run of the density with $z$ are known, $K_z(z)$ can be determined by solving the integral equation (2.10). Many such determinations of $K_z(z)$ have been made since Oort, with often controversial results. In particular, it is difficult to separate the vertical motion from the lateral one for stars far from the plane. Also, the same is true close to a spiral arm whose mass perturbs the gravitational field. We will only mention here a determination that rests entirely on the HIPPARCOS results. It is due to Michel Crézé and to his colleagues at the Strasbourg observatory. This is not the most recent one, but it has the merit of simplicity. It rests upon a complete sample of about 3 000 A stars closer than 125 parsecs, with well-determined distances. The distribution of their vertical velocity is determined from the proper motion along $z$ of the stars at low galactic latitudes. The gravitational potential of the disk is assumed to be of the form $\Phi(z) = \alpha z^2$, a sufficient approximation for the small values of $z$ of the sample. The restoring force is then $K_z(z) = d\Phi(z)/dz = 2\alpha z$ in absolute value, hence the local density $\rho_0 = \alpha/2\pi G$ from equation (2.1). One obtains numerically:

$$\rho_0 = 0.076 \pm 0.015 \text{ M}_\odot\text{pc}^{-3}. \qquad (2.11)$$

Let us compare this "dynamical" value with the direct estimate of the densities of the different components of the Galaxy near the Sun. Star counts in the HR diagram of Figure 2.1 (right) together with an estimate of their mass give the local stellar density, to which one should add that of the fainter dwarf stars observed with the Hubble space telescope: Figure 2.4 shows the mass function of these stars. The resulting stellar mass density is 0.033 ± 0.003 $M_\odot$ pc$^{-3}$. The mass density of the stellar remnants, white dwarfs and neutron stars, is estimated as 0.015 $M_\odot$ pc$^{-3}$. That of planets around stars is negligible. Finally, the density of the interstellar medium is quite faint in the considered volume, because we are located inside an almost empty bubble of 80 pc radius (see later). Moreover, the average density within 1 kpc from the Sun is low: this is the radius of the so-called Gould Belt, which is a rather dense ring. The actual density of the interstellar matter is certainly lower than the average density of the disk at the Solar radius, which is about 0.04 $M_\odot$ pc$^{-3}$. Hence the local total density is probably of the order of 0.05 $M_\odot$ pc$^{-3}$. If the difference with the dynamical value is significant, it might be due to dark matter. However, all these determinations are uncertain and any serious claim in favor of a real difference is premature. GAIA should do much better, and will probably give a more credible answer to the question.

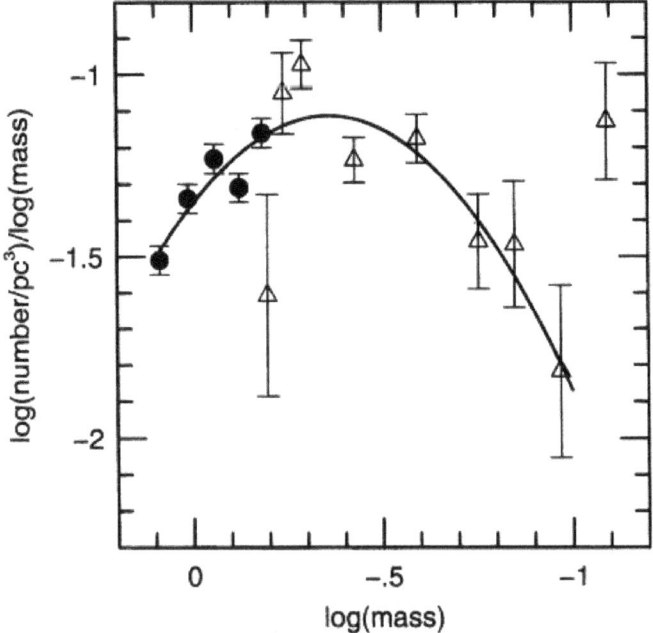

FIG. 2.4 – The mass function for dwarf, main sequence stars. The dark circles correspond to classical star counts, and the triangles with error bars from observations with the Hubble space telescope. From Gould, A., Bahcall, J.N. & Flynn, C. (1996), *Astrophysical Journal* 405, 759-768, with permission of the American Astronomical Society.

*The solar neighborhood*

The star counts also allow us to gain an idea of the thickness of the stellar Galactic disk. This can be derived from equation (2.7). For thin stellar systems, which correspond to a small velocity dispersion, we can assume $K_z(z) = -2\alpha z$, so that:

$$n(z) = n(0) \exp - (\alpha z^2/\sigma^2). \tag{2.12}$$

The scale height is in this case $z_0 = \sigma/(2\alpha)^{1/2}$. For thicker systems we should use a more realistic expression for $K_z(z)$.

The vertical velocity dispersion varies with the spectral type or the color of stars. For O and B stars, the most massive and hot, which are all young because of their small lifetime on the main sequence, it is 6 km/s. The velocity dispersion increases to 8 km/s for the A stars, to 11 km/s for the F stars, and to 21 km/s for the G to M dwarfs which are very old on average. It is 16 km/s for the red giants that have a variety of ages. The corresponding height scale varies from 80 pc for the youngest stars to 300 pc for the oldest ones in the disk.

## 2.3 Kinematics and dynamics of the stars of the local disk

The data from HIPPARCOS have allowed new studies of stellar kinematics on unbiased samples within a radius of about 100 pc around the Sun. Figure 2.5 shows the result of one of these studies; it gives the mean velocity of main sequence stars in the three orthogonal Galactic coordinates. While there is essentially no variation of the mean velocity with color in the direction of the Galactic center and perpendicular to the Galactic plane, there is a large variation in the direction of rotation. As to the velocity dispersion, it increases strongly in all directions for redder stars. All these quantities reach a fixed value for stars redder than $B-V = 0.61$: this color corresponds to the *Parenago discontinuity*, taken from the name of the Soviet astronomer who discovered this property. It is certainly an age effect, since the main sequence stars have an average age that increases with redder color: all the stars bluer than $B-V = 0.61$ are younger than 10 billion years old, while the reddest thus oldest stars on the average have an age of about 13 billion years. The "stellar gas" has been 'heated" during time by the encounters with density waves and interstellar clouds.

We can also see on Figure 2.5 that the old stars rotate less rapidly around the Galactic center than the younger stars, the difference being about 15 km/s. This phenomenon, called the asymmetric current, is explained as follows: the stars follow an epicyclical orbit around the Galactic center, which is the combination of rotation and of an approximately elliptic motion. These orbits depart more from a circle if the velocity dispersion is larger. A star close to the Sun that seems late with respect to pure rotation is actually spending most of its time inside the Solar circle. Due to the differential rotation, its mean rotation velocity is lower than that of the Sun. Conversely, a star that spends most of its time outside the solar circle is early with respect to pure rotation at this circle. Because the stellar density decreases strongly towards the outer Galaxy, there are more late stars than early stars, as can be seen in Figure 2.5. The effect, hence V, is larger when the velocity dispersion is larger, or the age older.

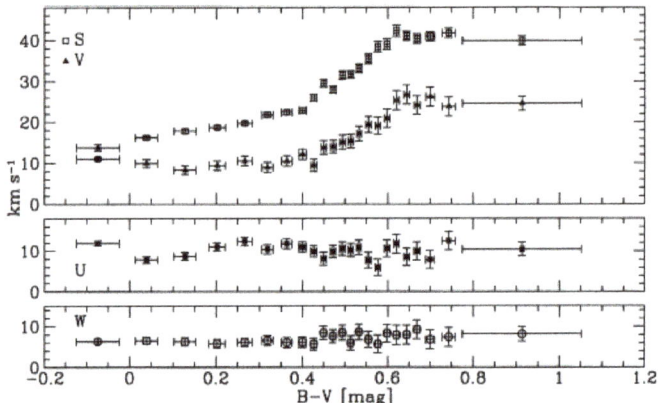

FIG. 2.5 – The three components of the mean velocity of the disk main sequence stars, as a function of their $B$-$V$ color. $U$ is the component towards the Galactic center, $V$ that in the direction of the rotation and $W$ that towards the Galatic North. The full circles at the top show the variation with color of the velocity dispersion in the direction of rotation. From Dehnen, W. & Binney, J.J. (1998) *Monthly Notices of the Royal Astronomical Society* 298, 387-394, Wiley, with the permission of the Editor.

To obtain the circular velocity at the solar circle, one has to extrapolate the relation between V and the velocity dispersion $\sigma_V$ to a null dispersion; One gets in this way the particular velocity of the Sun with respect to a point at the same position that would have a pure rotation around the Galactic center. This point is called the *local standard of rest* (LSR). Recent values for the three components of the velocity of the Sun with respect to the LSR are:

$$U_\odot = 11.1 \pm 0.7 \text{ km/s};$$
$$V_\odot = 12.2 \pm 0.5 \text{ km/s};$$
$$W_\odot = 7.3 \pm 0.4 \text{ km/s},$$

corresponding to a velocity of 18.0 km/s towards the apex with galactic longitude $l = 47.7°$ and galactic latitude $b = 23.9°$, in the Hercules constellation. These velocity components are different from those that were recommended

*The solar neighborhood*

before HIPPARCOS: at that time one distinguished the *kinematic velocity* with respect to the neighboring stars (components 11.1, 16.6 and 7.9 km/s, for a velocity of 20 km/s towards $l = 56.01°$, $b = 23.16°$) from the *dynamical velocity* with respect to a point with pure rotation (components 9, 12 and 7 km/s, for a velocity of 16.6 km/s towards $l = 53.28°$, $b = 24.60°$); The recent values given above correspond to the dynamical velocity.

The study of the proper motions and of the radial velocities of young stars as a function of their distance allowed Oort to give the parameters of the differential rotation near the Sun, written as the *Oort constants* A and B:

$$A = 1/2 \ (\Theta/R - \mathrm{d}\Theta/\mathrm{d}R)_\odot = -1/2 \ (R\mathrm{d}\Omega/\mathrm{d}R)_\odot, \text{ and} \quad (2.13)$$

$$B = A - (\Theta/R)_\odot = -1/2 \ (\Theta/R + \mathrm{d}\Theta/\mathrm{d}R)_\odot = -(\Omega + 1/2 \ R\mathrm{d}\Omega/\mathrm{d}R)_\odot. \quad (2.14)$$

$\Theta$ being the linear rotation velocity, $\Omega$ the angular velocity and $R$ the distance to the Galactic center. The $\odot$ symbol means that the quantities are taken at the level of the Sun. It is easy to see that the radial velocity $v_{\mathrm{rad}}$ and the proper motion $\mu$ of a star at the distance $r$ from the Sun and with galactic longitude $l$ are given to a first approximation by:

$$v_{\mathrm{rad}} = rA \ \sin 2l, \text{ et} \quad (2.15)$$

$$\mu = A \cos 2l + B \quad (2.16)$$

The present values of the Oort constants, derived from the HIPPARCOS observations, are:

$$A = 14.8 \pm 0.8 \text{ km s}^{-1} \text{ kpc}^{-1}, \text{ and} \quad (2.17)$$

$$B = -12.4 \pm 0.6 \text{ km s}^{-1} \text{ kpc}^{-1}. \quad (2.18)$$

We have seen that the velocity dispersion of stars contains important information on the dynamics of stars. This dispersion is not isotropic. Detailed study shows that it is also not symmetric with respect of the direction of the Galactic center or of the rotation. It follows, as a first approximation, a velocity ellipsoid represented on Figure 2.6. Its principal axis deviates from the direction of the Galactic center by the galactic longitude $l_v$, which depends on the color of the stars. This is called the *deviation of vertex*. It is of the order of 30° for young stars and decreases to about 10° for the old, red stars, whilst the velocity dispersion along the principal axis $V_1$ increases from 14 to 38 km/s, and proportionally on the two other axes. The main cause for this deviation is probably an asymmetry in the gravitational potential of the Galaxy, due to the spiral arms, which mainly affects newly born stars. However, a detailed study of the stellar velocities near the Sun shows that particular motions exist corresponding to stellar currents. They are identified on Figure 2.7. The average velocity of these groups of stars reflects the kinematic conditions at their formation. They contribute to the deviation of vertex.

## E2.1. Galactic coordinates

Astronomers use a system of galactic coordinates defined in Figure E2.1. To transform the galactic coordinates into celestial coordinates, right ascension and declination, one can use the NED calculator: http://ned.ipac.caltech.edu.

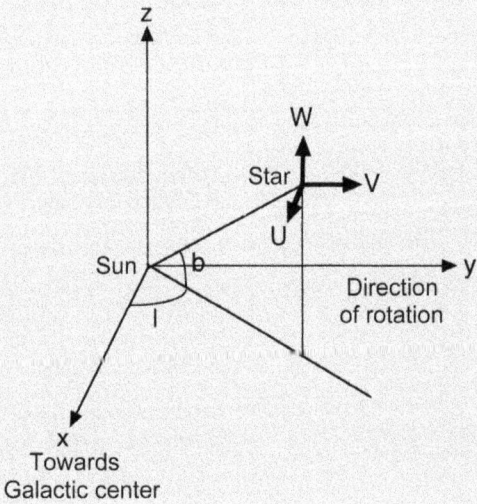

FIG. E2.1 – Galactic coordinates. The direction of a celestial body is defined by the longitude $l$, counted in the Galactic plane from the direction of the Galactic center (in constellation Sagittarius) in the trigonometric sense, and by the latitude $b$ counted positively towards the north. The vectors **U**, **V** and **W**, parallel to the directions $x$, $y$ and $z$, indicate the velocity of the star.

The solar neighborhood

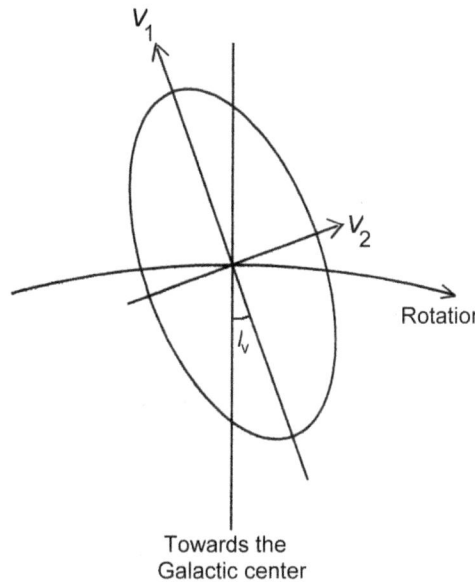

FIG. 2.6 – Cut of the velocity ellipsoid in the plane of the Galaxy. The velocity dispersion in any direction is proportional to the length of the line radius vector. The direction $V_1$ is that of the principal axis. $l_V$ is the galactic longitude of the deviation of the vertex.

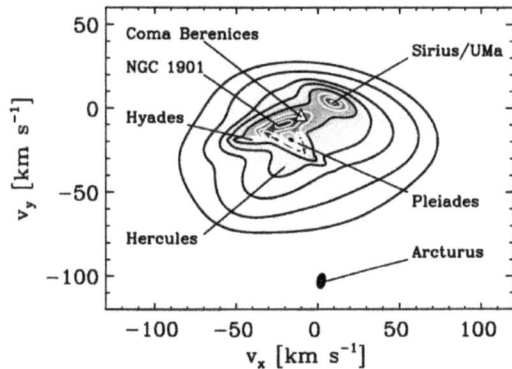

FIG. 2.7 – The stellar currents near the Sun, based on an unbiased sample of nearly 12 000 main sequence stars observed by HIPPARCOS. The density of stars projected on the Galactic plane is represented as a function of their velocity. The solar velocity is the reference (coordinates 0, 0) and the LSR velocity is indicated by a triangle. The names of the currents are indicated. From Bovy, J., Hogg, D.W. & Roweis, S.T. (2009) *Astrophysical Journal* 700, 1794-1819, with the permission of the American Astronomical Society.

## 2.4 High-velocity stars

Some stars have very large radial velocities and/or proper motions and have therefore a high velocity with respect to the Sun or the LSR. Some of them are massive stars that were part of a binary set whose other component exploded as a supernova. They were ejected with high orbital velocity. These are called the *run-away stars*. Other high-velocity stars, far more numerous, have a low mass. Two examples of orbits of such stars are displayed on Figure 2.8.

These stars form a system whose average rotation around the Galactic center is practically null. They have low heavy element abundances. Figure 2.9 shows the relation between this abundance and rotation.

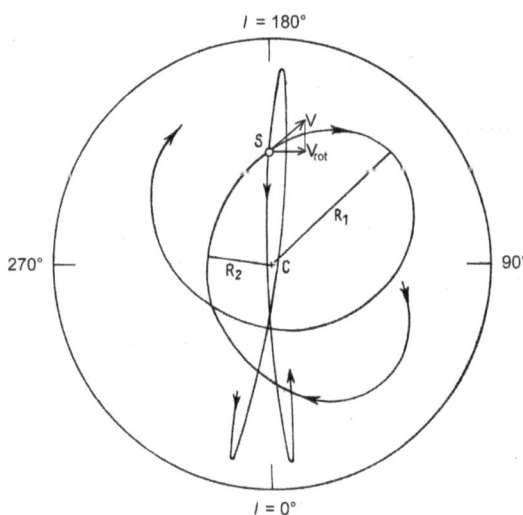

FIG. 2.8 – Two orbits of high-velocity low-mass stars, projected on the Galactic plane. C is the Galactic center and S is the Sun. They are presently near the Sun with a velocity known in magnitude and direction. Their orbit is calculated from this and a model of Galactic gravitational potential. $R_1$ and $R_2$ are respectively their maximum distance from the Galactic center (the *apogalacticum*) and their minimum distance (*perigalacticum*). The motion of the star along the less eccentric orbit is in the same sense as the Galactic rotation; the other orbit is retrograde. By definition, the rotation velocity $V_{rot}$ of the star is the projection of its space velocity on the direction of the Galactic rotation. For the Sun it is about 220 km/s, but it is much smaller, and often negative, for stars with very eccentric velocity, which belong to the Galactic halo. From Eggen, O.J., Lynden-Bell, D. & Sandage, A.R. (1962) *Astrophysical Journal* 136, 748-766, with permission of the American Astronomical Society.

The solar neighborhood

Actually, most of these high-velocity stars also have a high velocity perpendicular to the Galactic plane, and form an approximately symmetrical system around the Galactic center: this is the Galactic halo, which is also characterized by globular clusters (see Fig. 1.3). Many of these stars have escaped from the globular clusters, which have similar kinematics. We will come back to this in the following chapters.

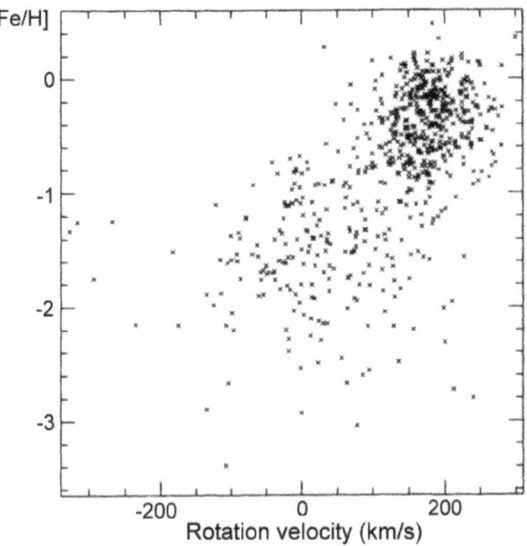

FIG. 2.9 – Relation between the abundance of heavy elements of stars near the Sun, represented by the quantity [Fe/H] = log(Fe/H) − log(Fe/H)$_\odot$, and their rotation velocity around the Galactic center (see its definition Fig. 2.8). The stars that rotate like the Sun with a velocity of the order of 200 km/s have relatively high abundances, while the high-velocity stars that rotate slowly or have a retrograde orbit ($V_{rot} < 0$), have an abundance of iron very much smaller, with a very large dispersion. Some have an abundance less than 1000 times that of the Sun. From Nissen, P.E. & Schuster, W.J. (1991) *Astronomy & Astrophysics* 251, 457-468, with permission of l'ESO.

## 2.5 The interstellar matter near the Sun

In order to compare the local interstellar matter with that in the whole Galaxy, it is of interest to look at Table 2.1, which summarizes the properties of its various phases.

The Sun emits a hot ionized gas whose density at the Earth's orbit is approximately 10 particles per cm$^3$ and the velocity is 400 km s$^{-1}$. If there were only vacuum in the Solar neighborhood, this solar wind would extend to infinity. However, there is some gas outside. When the momentum of the gas of the wind per unit volume equals the pressure of the external gas, a shock forms that limits the heliosphere, i.e. the cavity dug by the solar wind. This limit has been crossed by both Voyage probes, on 16 December 2004 at 94 astronomical units from the Sun for Voyager 1, and on 30 August 2007 at 84 u.a. by Voyager 2.

The ULYSSES probe and others have observed helium atoms from the interstellar medium, which had crossed the shock without being perturbed. After small corrections, it is possible to determine their properties in the surrounding interstellar gas: their density is $n_{He} = 0.015$ atom cm$^{-3}$, their temperature $T = 6300$ K and their arrival velocity with respect to the Sun $v = 26$ km/s. They come from the galactic longitude $l = 3.7°$ and the galactic latitude $b = 15.3°$. This velocity and this direction differ from those of the apex ($v = 13.4$ km/s, $l = 27.47°$ et $b = 32.57°$), showing that the local interstellar medium moves with respect to the LSR. From these data and some complementary data, one derives the atomic hydrogen density in the surrounding interstellar medium, $n_H = 0.19$ cm$^{-3}$, and that of the hydrogen ions, $n_{H+} = 0.06$ cm$^{-3}$. This low-density, warm and partly ionized medium is characteristic of the warm HI medium of Table 2.1. Observations with ULYSSES have also shown the arrival of interstellar grains that penetrate the heliosphere with the same velocity and from the same direction as the gas.

Further away from the Sun the situation becomes complex, testifying for the extreme inhomogeneity of the interstellar medium in general. The most used method to study its distribution, kinematics and physics consists in observing the absorption lines produced in the spectrum of stars

TAB. 2.1 : The different phases of interstellar matter in the Galaxy. Note that the warm atomic medium is partly ionized. The distribution of the atomic medium between the cold and the warm phases is somewhat problematic. The total amount of molecular gas is uncertain.

| Medium | | Density (cm$^{-3}$) | Temperature (K) | Total mass (M$_\odot$) |
|---|---|---|---|---|
| Atomic (HI) | Cold | ≈ 25 | ≈ 100 | $4 \times 10^9$ |
| | Warm | ≈ 0.25 | ≈ 8 000 | $4 \times 10^9$ |
| Molecular (H$_2$) | | ≈ 1 000 | ≤ 100 | $\geq 3 \times 10^9$ |
| Ionized | HII regions | ≈ 1 – 10$^4$ | ≈ 10 000 | $5 \times 10^7$ |
| | Diffuse | ≈ 0.03 | ≈ 8 000 | $10^9$ ? |
| | Hot | ≈ 6 × 10$^{-3}$ | ≈ 5 × 10$^5$ | $10^8$ ? |

by intervening atoms, ions or molecules. As the distance of these stars is known, this gives an upper limit of the distance of the gas detected in this way. The lines are in the visible the D lines of neutral sodium at 588.9 and 589.5 nm (Fig. 2.10), and the K lines of ionized calcium at 393.3 and 396.8 nm. Presently, the GAIA satellite is detecting an absorption band probably due to complex interstellar molecules in the spectrum of many stars. In the far ultraviolet observed by space telescopes (the Hubble and the FUSE ones), many lines have been observed, including those of multi-ionized species as $C^{3+}$, $N^{4+}$ and $O^{5+}$ (CIV, NV and OVI in astronomical notation), which characterize very hot gas. This gas also produces soft X rays that have been observed with space vehicles. The 21-cm line gives information on the HI medium. One can also measure the reddening of the light of stars, which gives information on the amount of intervening dust, (generally well mixed with the gas).

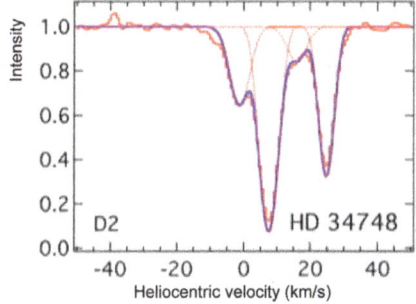

FIG. 2.10 – An example of an interstellar absorption line (the $D_2$ line of sodium) in the spectrum of a nearby star, which is at a distance of about 220 pc in the direction $l = 203.3°$, $b = -21.0°$. It is decomposed into four distinct Gaussian components. From Welsh, B.Y., Lallement, R., Vergely, J.a-L. & Raimond, S. (2010) *Astronomy & Astrophysics* 510, A54, with permission of ESO.$

Figure 2.11 shows the result of a recent study of the local interstellar matter, based on sodium absorption lines. It is perfectly confirmed by the study of the reddening by interstellar dust, and it was indeed long suspected, although it was impossible to give figures. We notice that the region within about 80 pc from the Sun is essentially empty, with the exception of a wall at 15 pc which appears to separate two regions, and that the matter further away is distributed in irregular clouds. These clouds are also observed in the 21-cm line and, when molecular, in the CO lines. The molecular clouds being more or less opaque and giving saturated absorption lines, it is difficult to see what is behind.

There is however some tenuous gas in the hole, even if it does not produce detectable absorption lines in the visible. It produces a soft X-ray emission and gives absorption lines of multi-ionized ions in the far-UV spectrum of B

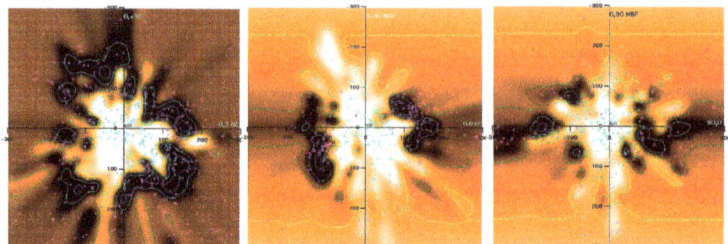

FIG. 2.11 – The local interstellar medium, mapped from the interstellar sodium absorption lines observed in front of the stars indicated by green or red triangles. Each square is centered on the Sun and measures 600 x 600 pc. From left to right: projection on the Galactic plane, the direction of the Galactic center being to the right; projection on a plane perpendicular to the Galactic plane and containing the Galactic center (to the right), and projection on a plane perpendicular to the Galactic plane and to the direction of the Galactic center. Absorption is not detectable in the white regions, and is increasingly strongly from yellow to black. The density in the most external regions is uncertain, as well as that in the black regions with matrices of white points. From Welsh, B.Y., Lallement, R., Vergely, J.-L. & Raimond, S. (2010) *Astronomy & Astrophysics* 510, A54, with permission of ESO.

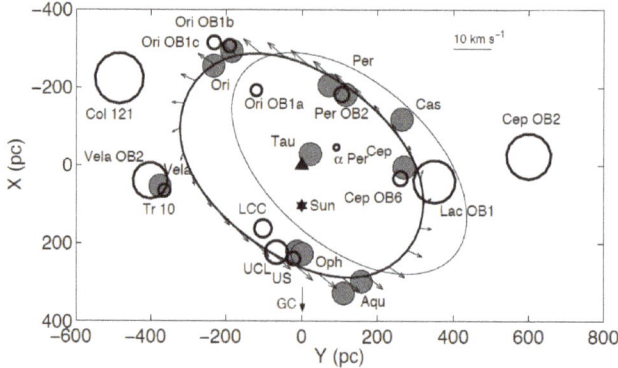

FIG. 2.12 – The Gould belt. It is schematized by a thick ellipse in projection on the Galactic plane, and its kinematics by arrows. The thin ellipse is a previous determination based on HI observations alone. The empty circles are O-B associations with their approximate size, and the grey circles symbolize molecular clouds. The center of the belt is noted by a triangle, and the position of the Sun is indicated. The oblique line is the intersection of the Galactic plane and of the plane of the belt, which is inclined by 17°. The molecular complexes closest to the Sun are those of Taurus and Ophiuchus, at 150 pc. From Perrot, C.A. & Grenier, I.A. (2003) *Astronomy & Astrophysics* 404, 519-531, with permission of ESO.

*The solar neighborhood*

stars of its periphery. This is clearly an interstellar bubble, perhaps double, which was dug by the explosion of several supernovae, the most recent one being approximately 4 million years old. Some small warm clouds managed to survive inside this bubble: there are about fifteen of them within 15 pc from the Sun, including the one that contains the heliosphere. A very cold ($\approx$ 20 K) HI cloud is also present at a distance of about 20 pc, in the Leo constellation.

Further away from us, the average interstellar density remains small until we reach, at a few hundred parsecs, a dense annular structure inclined on the Galactic plane: this is the Gould belt described in 1874 by this American astronomer. It contains a large amount of molecular gas in which many stars have been formed and are still in formation (Fig. 2.12).

The exact origin of this structure is unknown, but it is clear that it is due to some violent event 50 to 100 million years old. There are similar structures in external galaxies.

# Chapter 3
# Structure and components of the Milky Way

In this chapter, we will deal with the Galaxy as a whole. The observation of the Galactic disk is much less easy than that of the Solar neighborhood, because we are located inside this disk in a rather eccentric position, and interstellar extinction prevents optical observations at distances larger than a few kiloparsecs. Fortunately, the infrared and radio waves are little or not absorbed, and have supplied, for the last few decades, much information on the remote regions of the disk. Moreover, the observation of external spiral galaxies is very helpful to understand by analogy the properties of our own Galaxy.

Here, we will not discuss the Galactic center, which will be the subject of the next chapter.

## 3.1 Dimensions and rotation of the Galaxy

In the introductory chapter, we recalled the beginnings of the determination of the dimensions and of the rotation of the Galactic disk. Until 1950, the rotation curve, which gives the rotational velocity as a function of the distance to the Galactic center, was unknown: only its slope near the Sun was known, under the form of the Oort's constants. Then, the rotation curve could be determined for the first time thanks to the 21-cm line of atomic hydrogen. Figure 3.1 shows the principle of this determination.

Let us assume that the spectrum of the 21-cm line is observed in the Galactic plane in the direction of the galactic longitude $l$. From each point of the line of sight, for example from point P, comes a component of the line, shifted by the Doppler-Fizeau effect corresponding to the relative velocity of this region with respect to that of the LSR. The radial velocity of P with respect to the LSR is:

$$v_{rad} = \Theta \sin(l+\beta) - \Theta_0 \sin l = R_0 (\Theta/R - \Theta_0/R_0)\sin l = R_0 (\omega - \omega_0)\sin l, \quad (3.1)$$

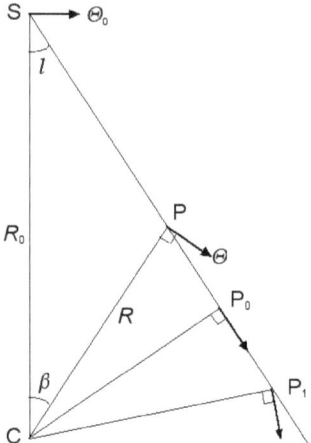

FIG. 3.1 – Determination of the Galactic rotation using the 21-cm line. C is the Galactic center, S the Sun and $\Theta_0$ the rotation velocity of the LSR. The rotation velocity at P is $\Theta$. Explanations in the text.

where $R_0$ is the distance of the Sun to the Galactic center and $\omega$ the angular velocity of rotation. Because $\omega$ generally increases with decreasing distance to the center, the highest velocity is that of point $P_0$. This opens the possibility of obtaining $\Theta R_0$ at different longitudes, from the maximum velocity of the different line profiles, hence the rotation curve $\Theta(R)$, once $R_0$ and $\Theta_0$ have been chosen. From the relations (2.13) and (2.14) of the preceding chapter, we see that the slope of the rotation curve near the Sun, as determined now by the stars and not the gas, is:

$$(d\Theta/dR)_0 = -(A+B) = -2.4 \pm 1.0 \text{ km s}^{-1} \text{ kpc}^{-1}, \qquad (3.2)$$

$A$ and $B$ being the Oort's constants. This slope is very small, so that the local rotation curve is practically flat near the Sun, in agreement with the 21-cm results.

However, the preceding method does not apply to the galactic radii larger than $R_0$. It is then necessary to use objects for which both the distance and the radial velocity are known. Such objects are cepheids, which are massive pulsating stars, HII regions whose distance is determined from the photometry of the ionizing stars, or molecular clouds forming stars whose photometric distance can also be obtained. However, only the 21-cm line can be used at very large distances that are then rather difficult to determine.

The rotation curve obtained in this way is displayed on Figure 3.2. It is roughly flat, but with important undulations that will be commented on later. A special determination using only cepheids gives a somewhat different rotation curve, which is practically flat up to a galactocentric radius of 15 kpc, beyond which the observations are not numerous enough (Fig. 3.3).

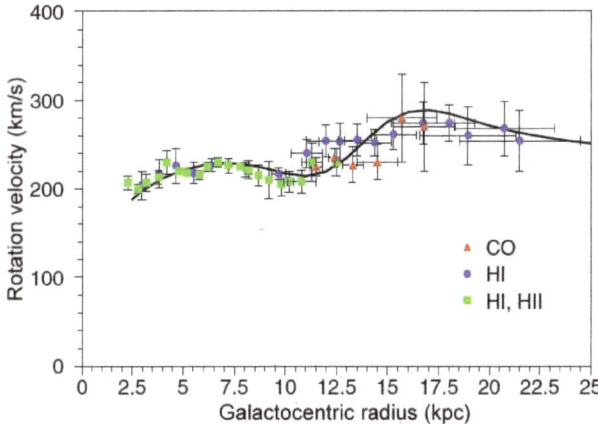

FIG. 3.2 – The rotation curve of the Galaxy obtained from observations in the 21-cm line of atomic hydrogen (HI), of HII regions and of molecular clouds in the CO line. The curve corresponds to a model. The Sun is supposed to be at 8.3 kpc from the Galactic center, and the rotation velocity at the solar radius is taken as 220 km/s. From de Boer, W. et al. (2005) *Astronomy & Astrophysics* 444, 51-67, with permission of ESO.

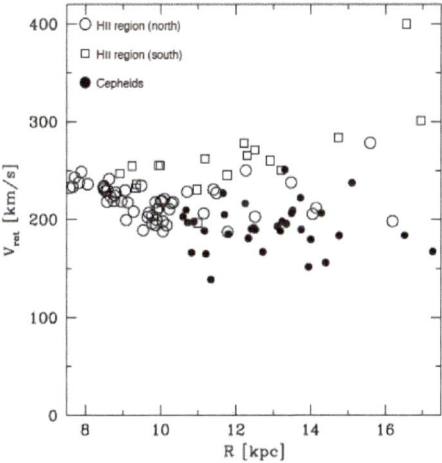

FIG. 3.3 – Comparison of the rotation velocities obtained from HII regions and from cepheids. The Sun is supposed to be at 8.5 kpc from the Galactic center, and the rotation velocity at the solar radius is taken as 200 km/s. These choices do not affect the result much. The high rotational velocities of the gas in the southern hemisphere are not confirmed by the cepheids. The difference might be due to asymmetric motions of the gas. From Pont, F. et al. (1997) *Astronomy & Astrophysics* 318, 416-428, with permission of ESO.

Once the rotation curve is determined, we can use the 21-cm line observations to obtain the distance of the emitting regions from their radial velocity. Figure 3.4 shows the distribution of the radial velocities in the Galactic disk, as seen from the LSR. For a given radial velocity, there is an ambiguity in the internal regions of the disk: the points P and $P_1$ of Figure 3.1, symmetrical with respect to $P_0$, have the same radial velocity. To solve this ambiguity, one may use the fact that the thickness of the gas disk is approximately uniform in the inner regions: if we observe at a sufficiently high galactic latitude, we see only the nearest regions. The same reasoning holds for the CO lines. In any case, the kinematic determination of the distances is very uncertain, not only because of the departures to pure rotation, but also because of the random motions of the interstellar medium. Moreover, the radial velocity is null in the direction of the center and of the anti-center, and small at galactic longitudes near 90° and 270°, at least for the closest regions. In the best conditions, the uncertainty on the distance is already of the order of 0.5 kpc. It is impossible to do better for the distribution of the atomic gas, so that the old map displayed on Figure 1.4 could only be marginally improved.

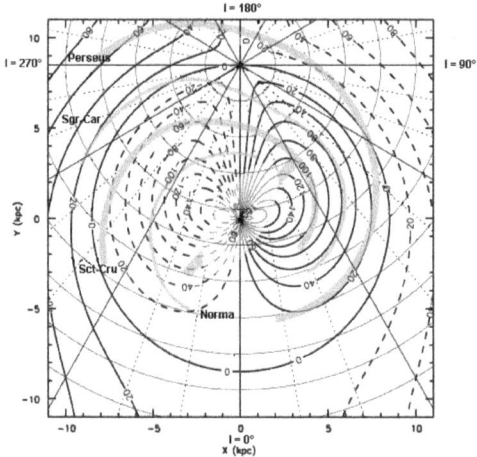

FIG. 3.4 – The radial velocity field in the Galactic disk as seen from the LSR. The Sun, hence the LSR, is at position x = 0, y = 8.5 kpc. The Galactic center is at position (0, 0). The lines of equal positive velocity (receding regions) are full, and those of equal negative velocity are dashed. The velocities are in km/s, and the rotation velocity at the Sun is taken as 220 km/s. The main spiral arms are schematized in grey, with their names. From Benjamin, R.A. (2008) *Astronomical Society of the Pacific Conference Series* **387**, 375-380, with permission of the American Astronomical Society.

The situation is better for HII regions and molecular clouds when their distance can be determined from that of associated stars. Figure 1.6 of the preceding chapter displays the distribution of HII regions with known distance. Some portions of spiral arms are well defined in the picture, but the fit with four logarithmic spiral arms as depicted in this Figure is somewhat problematic.

The most accurate distances in the Galaxy come from Very Long Baseline Interferometry (VLBI) in radio. For example, it is possible to obtain with remarkable accuracy the geometric parallax of some young, massive stars which emit radio waves, by measuring, at different times of the year, their position with respect to a reference system materialized by distant quasars. The sources studied most in VLBI are interstellar masers associated to stars in formation. Some very young stars expel with a high velocity many compact masses of molecular gas that emit, as natural masers, intense and very narrow radio lines of some molecules: $H_2O$, SiO and methanol $CH_3OH$. Their lateral motion (proper motion) can be followed by VLBI observations at different times. The statistical combination of these proper motions with the radial velocities that are also obtained very accurately gives the distance, assuming isotropy of the ejections, a hypothesis that can be checked *a posteriori*. The accuracy on the distances determined in this way is of the order of 1% at 1 kpc and 10% or better at 10 kpc. The distance to the Galactic center has been obtained recently in this way as 8.3 ± 0.2 kpc and the rotation velocity of the LSR as 246 ± 7 km/s, instead of the canonical 8.5 kpc and 220 km/s (the convention of the IAU International Astronomical Union).

In some cases, it is possible to obtain a rough idea of the distance of radiosources close to the Galactic plane by observing the 21-cm line in absorption in front of the radio continuum of these sources, and by comparing the absorption spectrum with the emission spectrum near the source: the components that are lacking in the absorption spectrum correspond to regions more distant than the source. An example can be found later in Figure 3.10. This method is the only one that can be used for most supernova remnants.

Detailed studies of the Galactic disk show that, setting aside the spiral structure, it does not possess symmetry of revolution, especially in its external parts. Various asymmetries and distortions are present in the velocity field as well as in the distribution of the gas. The most important distortion is the warping shown on Figure 3.5, which is much more important on one side of the Galaxy. It affects not only the neutral gas, but also the molecular one, the young stars and even a part of the old stars: it is thus certainly of gravitational origin. Such warps, which generally come with a thickening of the disk, are quite common in external galaxies, isolated or not. We will examine its causes in Chapter 8.

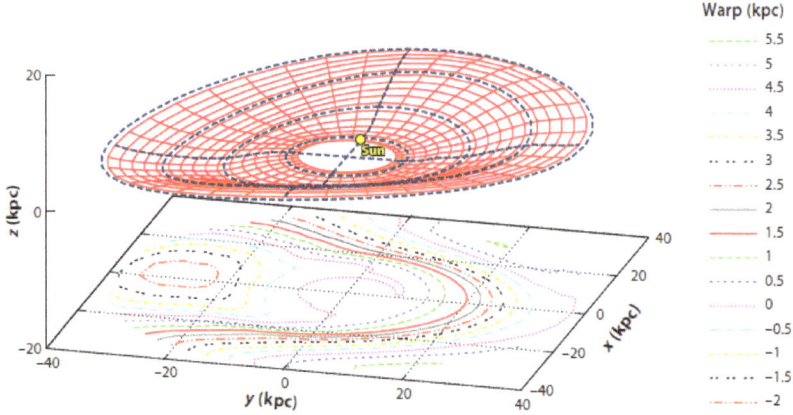

FIG. 3.5 – The warp of the Galactic disk. Bottom, map of the distance of the center of the vertical distribution of interstellar neutral hydrogen to the mean Galactic plane (scale on the right). Top, a 3-D view of the warp. The Galactic center is the origin of the coordinates, and the position of the Sun is indicated. Figure kindly communicated by Peter Kalberla.

It is possible from the rotation curve to obtain some information on the distribution of the mass in the Galaxy, ignoring these complications. One has to assume gravitational equilibrium, the centrifugal force being compensated by the gravity. On doing that, a difficulty immediately arises: the rotation curve is practically flat up to the largest radius where observations are possible, about 15 kpc from the center, while, because this region seems relatively devoid of stars and gas, we would think that the rotation velocity should decrease with increasing radius, following Kepler's law ($V$ proportional to $R^{-1/2}$). This phenomenon is absolutely general for spiral galaxies. To account for it, one generally assumes that a massive, more or less spherical, halo of dark matter that contains most of the mass of the galaxy exists around spiral galaxies and is more extended than the disk. Another possible explanation, which will be developed at the end of this chapter, is the possible existence of a supplementary term to Newton's universal gravitation law, which would become important at very large distances (MOdified Newtonian Dynamics, or MOND). For the moment, we will consider the first explanation with the unmodified gravity law, and see what the constraints on the mass of the Galaxy and the distribution of this mass are.

Let us assume in this framework that the mass is distributed in a flat disk and in a halo of dark matter. If the mass had a spherical distribution, the velocity $V(R)$ at the galactocentric radius $R$ would only depend on the mass $M(R)$ inside the radius $R$. From $V(R)$, one would derive $M(R)$ by the relation:

$$M(R) = R\ V^2(R)/G, \qquad (3.3)$$

where G is the constant of gravitation. Numerically, this gives:

$$M(R) = 2.325 \times 10^5\ R\ V^2(R), \qquad (3.4)$$

where $M(R)$ is expressed in solar masses $M_\odot$, $R$ in kpc and $V$ in km/s.

The case of a flat disk, the thickness of which we will neglect as indeed suggested by observation, is more complex because the mass exterior to radius R intervenes. $M(R)$ then depends on the velocity $V(r)$ at all radii $r$. In order to estimate $M(R)$ for our Galaxy, we have to consider the whole rotation curve. It is not defined in the central regions which are not axi-symmetric: for the sake of simplicity we will consider a rotation curve increasing linearly with radius from 0 at the center until some radius $a$. This hypothesis does not constrain the problem much as long as $a$ is not larger than a few kpc, which is the case for our Galaxy. Then we will assume a flat rotation curve until the radius where there are no more observations. Then we suppose either that the rotation velocity stays constant, or that there is no more mass so that the rotation curve is keplerian: $V(R) \propto R^{-1/2}$.

The result is given on Figure 3.6, where $M(R)$ is plotted for a flat disk as a function of what would be obtained for a spherical distribution of mass, giving the same velocity at the same radius. We see that the mass within a radius $R$ is between 1 and 0.6 times the value given by formula (3.3) or (3.4), according to the hypothesis made on the more or less spherical distribution of the dark matter. This uncertainty is not much larger than that due to our imperfect knowledge of the disk shape. We then obtain the following dynamical masses:

$$M(8.3\ \text{kpc}) \approx 0.75 \times 10^{11}\ M_\odot \qquad (3.6)$$

inside the Solar circle, with a rotational velocity of the LSR of 246 km/s, and assuming that all the mass is in the disk. If a part of the mass is in a central sphere, the total mass is slightly higher.

$$M(21\ \text{kpc}) = 2 - 3 \times 10^{11}\ M_\odot \text{ where the observations end.} \qquad (3.7)$$

The total mass of gas and stars in the Galaxy is estimated as $0.85 \times 10^{11}\ M_\odot$ (see Table 2.1 for the gas and Fig. 3.8 for the stars). Most of this matter is inside the solar circle. This value is close to that of equation (3.6), so that dark matter is not necessary in the internal regions of the Galaxy. Conversely, a large amount of dark matter seems necessary to account for the flat rotation curve beyond the Sun.

In the MOND models, the result is completely different as we will see later.

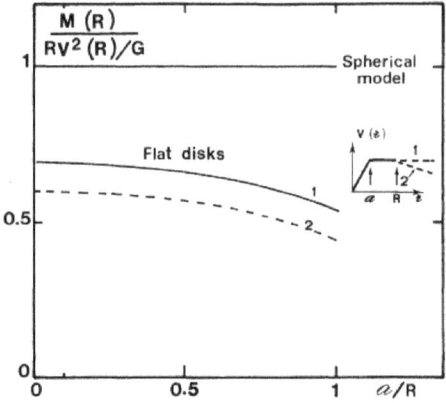

FIG. 3.6 – The mass $M(R)$ inside radius $R$ for different models of mass distribution. It is relative to the mass of a spherical model. The disk model 1 has a flat rotational curve from radius $a$ to far beyond the radius $R$. The disk model 2 has no mass beyond R, hence a keplerian rotation there. Any galaxy with cylindrical symmetry has a mass $M(R)$ inside radius $R$ between that of model 2 and that of the spherical model. From Lequeux, J. (1983) *Astronomy & Astrophysics* 125, 394-395, with permission of ESO.

## 3.2 Stellar populations in the Galaxy

The Milky Way is a late-type spiral galaxy, which is still forming stars at a rate of a few per year. Its type is more precisely Sbc, i.e. with an average bulge. We have known for several decades that it has also a bar of medium strength, so its type is rather SBbc or SABbc. The image of our Galaxy obtained in the near infrared by the COBE satellite, and then the large 2MASS infrared star survey, have indeed shown the presence of the bar and of a small bulge (see Fig. 1.8). The latter has a somewhat square shape, like a box, and even has an extension on each side, like a peanut. We will see in Chapter 5 how these box or peanut-shaped bulges form from the stellar bar, which develops as a gravitational instability of the galaxy disk. The bar instability is a density wave which rotates in the disk, and is accompanied by spiral waves, which rotate with a speed either equal to that of the bar or different. The rotational velocity of the wave is generally different from that of the stars in the disk, so that the stars periodically cross the bar or the spiral arms.

With regards to the stellar component, several structures can be distinguished: the bulge, the thin disk, the thick disk, and the stellar halo. The thin disk is the main component, the most massive. Of course it is not easy to study, because the Sun is part of this disk, which we thus see edge-on from the inside. In the visible light range, we mostly see the disk of dust that absorbs and reddens the light of the background stars, and it is only in the near infrared that we can see the entire thin disk (with still some extinction towards the center of the Galaxy). The thin disk contains young stars, and the interstellar medium from which new stars are formed. The other components are mostly made of old stars.

We can thus attempt to define stellar populations according to their ages. Historically, the German astronomer Walter Baade (1893-1960) who, in the US in 1944, observed the stars of the Andromeda Galaxy and classified its stars into two populations: Population I, composed of the young disk stars, and Population II made of the old stars of the halo. These two classes were also identified in our Galaxy. In the 1950s, studies of chemical abundances in stars showed that the stars of the disk (Population I), characterized by their dynamics (rotating disk), morphology (flattened disk) and young age were richer in heavy elements. Probably formed somewhat late in the evolution of the Galaxy, they took advantage of the recycling of heavy elements synthesized in the previous stars: indeed, throughout their lifetime, stars lose mass, and throw part of their gas into the interstellar medium, enriched by nuclear reactions within them. The halo stars (Population II) were believed to have been formed first, making a more or less spherical halo with no global rotation (see Section 2.4 above). This halo was then generally regarded as the remains of the beginning of the formation of the Galaxy. Their low abundance of heavy elements is consistent with this idea.

However, in the 1970-80s, increasingly numerous and accurate data on the dynamics and abundance of stars showed that the situation was much more complex. In all the different structures of the Galaxy, there is a mixture of abundances and ages, with a wide dispersion. In the bulge, which was initially supposed to contain only a relatively old Population II, there are also stars highly enriched in heavy elements, and young stars. Moreover, in the mid 1980s, a thick disk was discovered in our Galaxy, and also in a large number of external spiral galaxies. These thick discs have a radial distribution and rotation similar to the thin Population I disk, although their stars are old, with abundances approaching that of the Population II stars. We note for completeness that another population called Population III corresponds to stars formed in the Universe when it was only a tenth of its present age, or less. These first stars were devoid of heavy elements, and had a huge mass compared to the stars today. As the lifetime of such high-mass stars was very short, they have all disappeared, after having synthetized the first heavy elements in the Universe.

Given this mix of properties of stars in the different structures of the galaxy, the terms of Population I and Population II have become almost meaningless. We will now describe in some detail the four components of stellar populations: the halo, the bulge and the thin and thick disks, presented in turn.

### 3.2.1 The stellar halo

The stellar halo is the component of the Milky Way that contains less stars, a few percent of all stars in the Galaxy. It is characterized by several features: first its morphology, which is at the origin of its name as it forms a nearly spherical ensemble around the Galaxy center. The halo is larger than the disc, extending at least to 80 kpc from the center. Kinematically, the halo differs from the disk and the bulge because it does not rotate: its stars have disordered motions on orbits of all orientations, often highly elliptical. The halo stars are old (more than 12 billion years) and metal-poor, with an average abundance of iron about 30 times less than in the Sun with a large dispersion.

The stellar halo is not a homogeneous component. A fine structure was discovered there about fifteen years ago: in deep star counts, and especially using the color-magnitude diagrams that allow the various types of stars to be distinguished, stellar streams can be seen that extend as separate entities over hundreds of degrees in the sky (Fig. 3.7).

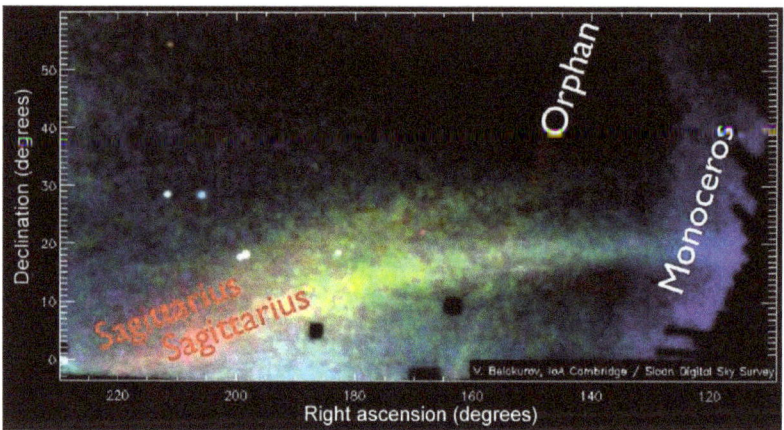

FIG. 3.7 – Map of the stellar density in the direction of the Galaxy anti-center (side opposite to the galactic center), at high galactic latitude: the middle of the picture is at 60° above the plane, which is to the right of the image. Three stellar currents (Sagittarius, Orphan and Monoceros) are visible in this image, the most important being that of Sagittarius which splits into two streams, corresponding to two portions of the orbit of a dwarf galaxy captured by our Galaxy, whose debris produce these streams. From Belokurov, V. et al. (2006) *Astrophysical Journal* 642, L137-140, updated version, with permission of the American Astronomical Society.

Each of these currents is well-defined in terms of position, velocity, and also age and metallicity of stars that constitute them. They probably come from the capture of small dwarf galaxies, former companions of the Milky Way. In the case of the Sagittarius dwarf galaxy that is still being under capture, what we see are two tidal tails produced by gravitational interaction with our Galaxy. In this case, one could reconstruct the orbit of the Sagittarius dwarf galaxy, along which the tidal debris are located. Its orbit made several turns around the Galaxy, like a rosette, with an amplitude ranging from 15 to 60 kpc. Debris then ended up in streams as shown in Figure 3.7. Such debris are found in various rings around the Galaxy, and it can be understood that a multitude of these currents could ultimately form the stellar halo around the Milky Way. It is not yet clear what is the relative share of pre-existing stars and of stars captured in the stellar halo, but some astronomers suggest that the halo could have been entirely formed by the remnants of dwarf galaxies that have fallen on the Milky Way since its formation. Because these kinds of galaxies have a very low surface brightness, and do not have gas, it is very difficult to discover new ones around the Milky Way. In spite of this, no less than 29 of these galaxies are presently known, thanks in particular to large sky surveys like the SDSS (Sloan Digital Sky Survey), and many others are still to be found. Note that two satellites of our Galaxy are more massive and gas-rich than these dwarf spheroidal galaxies: these are the two Magellanic Clouds, which are also in tidal interaction with the Milky Way. The resulting tidal tail of this interaction is the Magellanic stream, a polar ring around the halo of the Galaxy. This will be detailed in Chapter 8.

For completeness, we should say a few words about a stellar component whose morphology and dynamics approach the stellar halo: this is the ensemble of the globular clusters. These are very dense star clusters, containing typically 100,000 stars of homogeneous age. Their spherical distribution extends further than the stellar halo, up to 100 kpc from the galactic center. In the Milky Way, there are approximately 150 such clusters. Their stars are very old and very poor in heavy elements; this is not a general feature of galaxies, because in many of them globular clusters are young. We believe that these globular clusters were mostly formed during the merging of galaxies.

## 3.2.2 The bulge

The Milky Way has a modest stellar bulge. Its characteristics are intermediate between those of a "classical" bulge, i.e. a very concentrated ellipsoid without rotation, and those of a "pseudo-bulge", which is more similar to a disc because of its rotation, flattening, and concentration. We will see in following chapters how the action of the bar can form a pseudo-bulge from a disk by secular evolution. The morphology of the bulge, in the shape of a box or a peanut, reveals that the bar has contributed to its formation. Several types of stars accumulate in the bulge: most of them are old, but young stars are also present. The heavy-element abundance in the bulge stars has a wide dispersion, from very metal-poor to very rich. It is likely that several formation mechanisms have contributed to the development of the bulge, and current studies attempt to quantify how much of the mass is due to the internal evolution from the disk, what fraction is due to the gravitational action of companions, and what portion would come from tidal debris. The degree of rotation of stars around the galactic center is crucial to this determination: it is still poorly known, due to distance ambiguities for the stars and extinction by interstellar dust. Proper motions that will be obtained soon by the GAIA astrometry satellite will certainly bring progress in this field.

### 3.2.3 The thick disk

Most spiral galaxies have a thick disk (Fig. 3.8).

The thick disk of our Galaxy was discovered in the 1980-90s, thanks to the study of stellar populations in the solar neighborhood. All stars of this disk are old, approximately 12 billion years. Their abundance in heavy

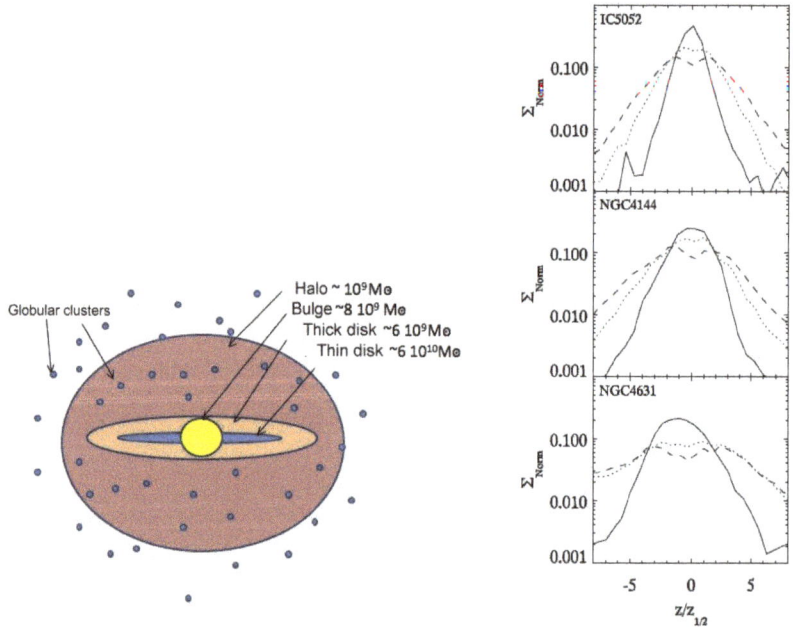

FIG. 3.8 – Left: Schematic view of the various stellar components of the Milky Way. Right: examples of thick disks in three galaxies seen edge-on. The curves show the normalized profiles of the logarithm of the surface density of young stars (solid line), of stars of intermediate age (dotted line) and of old stars (dashes). From Seth, A.C., Dalcanton, J.J. & de Jong, R.S. (2005) *Astronomical Journal* 130, 1574-1592, with permission of the American Astronomical Society.

elements is quite low: they contain 4 to 5 times less iron than the thin disk stars. But the distribution of abundances of the different elements is not the same: they are relatively richer in "alpha" elements such as oxygen or magnesium. These features are very useful in determining the mechanisms of formation, as will be discussed in detail in Chapter 6.

Like the thin disk, the thick disk is rotating, although more slowly, while the velocity dispersion of the stars is larger. Its thickness is about four times larger: 300 pc for the thin disk, 1.2 kpc for the thick disk at the Sun. In the outer parts of the Galaxy, both disks flare and their thickness increases. We also know that over time, the velocity dispersion of the stars in the thin disk increases, due to their deflection by gravitational potential accidents: giant molecular clouds and spiral arms. However, their height above the plane will never reach that of the thick disk stars. There are really two components, the formation mechanisms of which may be different. We will discuss them in chapters 5-8.

Thick discs are also seen in many spiral galaxies, when viewed edge-on. An example is shown in Figure 3.9.

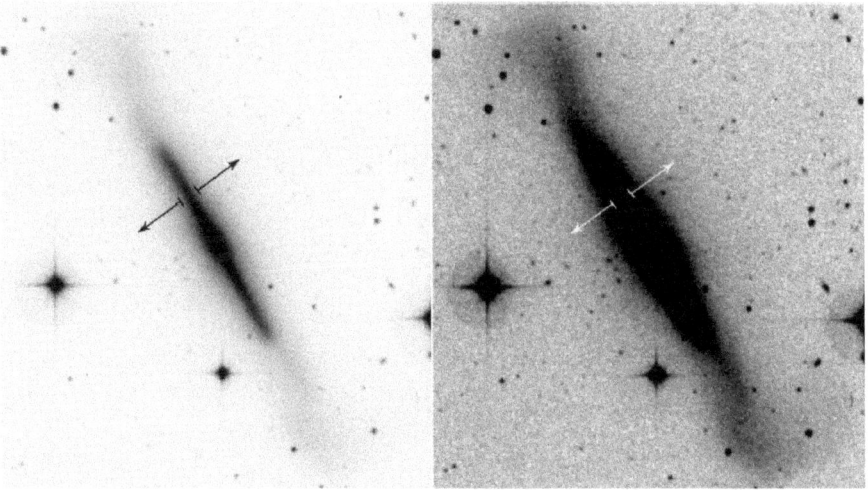

FIG. 3.9 – Negative images of the spiral galaxy NGC 4762. The arrows serve as markers to show the difference in thickness between the thin disk, seen with a low exposure (left) and the thick disk, seen with a deeper exposure (right). Note the flaring and largest thickness of these disks far from the center of the galaxy. From Tsikoudi, V. (1980), *Astrophysical Journal Supplement* 43, 365-377, with permission of the American Astronomical Society.

## 3.2.4 The thin disk

The thin disk is the main stellar component of the Galaxy. Its radial distribution obeys an exponential law, like all disks of spiral galaxies. Its characteristic radius (the one in which the density is divided by e = 2.72) is 3 kpc, but the disc extends to a galactocentric radius of 16 kpc at least.

The thin disk of stars is associated with an even thinner disk of gas, consisting mainly of neutral hydrogen, atomic or molecular according to its local density; we will see its properties later. New stars form from clouds of molecular gas at a rate of a few solar masses per year, and old stars return some gas to the disk when they die, so that the thin disk is constantly renewed. There is also an in-fall of extragalactic gas on the Galaxy, which settles in the thin disk. The young stars form in groups and open clusters of homogeneous age. But soon these clusters dissolve and their stars acquire disordered random velocities, due to encounters with spiral waves or with interstellar clouds and other clusters. The time required for a complete dissolution of dense clusters is of the order of a billion years (a few billions in rare cases); but the most massive stars, which live only a few million or tens of millions of years, have already disappeared. Moreover, the average life of the molecular clouds that give rise to clusters of stars is estimated as 40 million years: they are dispersed by stellar winds and supernova explosions.

Therefore, the younger stars are found mainly in the thin disk. But very old stars are present in the thin disk; their average age is estimated at some 8-10 billion years. There is a small abundance gradient in the thin disk, with larger abundance towards the center, as expected because star formation is more active in the central parts than in the outside. But this gradient is smaller than expected from the star formation rate in the different parts. In fact, the spiral waves and the bar which run continuously in the disk mix its matter radially. They are responsible for the radial migration of stars, which results in a radial mixing of ages and abundances. We will describe these processes in Chapter 5. The Sun, for example, today at about 8.3 kpc from the center, was probably not formed at this distance, but has probably migrated over time from more central regions of the Galaxy.

It is not easy to know the history of star formation in the Galaxy from the age distribution of its stars, which is quite uncertain. However, it has been possible to show that star formation has been relatively constant over time. This is contrary to what would be expected from a galaxy evolving as a close system, starting from pure gas whose fraction decreases progressively over time: then, the star formation rate being more or less proportional to the gas density, it should decrease exponentially with time. The more constant rate of star formation implies that the Galaxy has regularly accreted gas from the intergalactic medium.

## 3.3 The interstellar medium in the Galaxy

Appendix 2 recalls some basic notions concerning the interpretation of the spectral lines of the interstellar medium, and may be consulted with profit by readers unfamiliar with this topic, and also with the notations used by radio astronomers.

### 3.3.1 The atomic "neutral" medium

When the density is less than about 1000 atoms per cm$^3$, the interstellar hydrogen is substantially in the form of atoms when it is away from the ionizing UV radiation sources (photons of energy higher than 13.6 eV, or wavelength smaller than 91.1 nm). However, it remains bathed in stellar ultraviolet radiation of energy less than 13.6 eV, so that all atoms whose ionization potential is smaller than 13.6 eV are ionized: this is the case for all elements with the exception of noble gases, nitrogen and oxygen. However, the ionization potential of oxygen is so close to that of hydrogen that the charge exchange reaction $H^+ + O \leftrightarrow O^+ + H$ produces ionized oxygen when the temperature reaches a few hundred kelvin.

The 21-cm line provides the best tool to study the atomic neutral gas (Fig. 3.10). When the optical thickness in this line is low, which is often the case, we obtain the column density of atomic hydrogen by the simple expression:

$$N(\mathrm{H}) = 1.83 \times 10^{18} \int T_B dV \text{ atom cm}^{-2} \text{ (K km/s)}^{-1}, \qquad (3.8)$$

where $\int T_B dV$ is the integrated intensity of the line. According to common practice in radio astronomy, the intensity is expressed as a brightness temperature $T_B$ (see Appendix 2), and the frequency is transformed into radial velocity by the Doppler relationship $(\nu - \nu_0)/\nu_0 = (V - V_0)/c$, $V_0$ being a reference velocity, generally that of the LSR.

We deduce from observations that there are two main phases in the atomic medium: a relatively dense and cold phase ($n \approx 25$ cm$^{-3}$, $T \approx 20\text{-}100$ K) and a low density, warm phase ($n \approx 0.25$ cm$^{-3}$, $T \approx 8000$ K), which have similar masses in the galactic disk, $1.5 \times 10^9$ M$_\odot$ each (see Table 2.4). The distribution of both phases is highly inhomogeneous, in clouds, layers or filaments. The first phase is heated by the photoelectric effect on interstellar dust grains: the UV photons extract electrons from grains. These electrons have a certain kinetic energy that they communicate to the ions, and then the ions to hydrogen atoms. The cold phase is cooled primarily by the emission of the ionized carbon C$^+$ line at 158 micrometers, which is excited by collisions with free electrons and incidentally with H atoms. The second phase is primarily heated by interstellar X-ray emission and cooled by emission of the same C$^+$ line, with the addition of the neutral oxygen line at 63 micrometers.

At the Sun, the half thickness of the dense and cold phase is 100 pc, and that of the warm phase 180 pc. Figure 3.11 shows the radial distribution of the atomic component (cold + warm) of interstellar matter. Note that the projected mass surface density of this component is rather constant, of the order of 10 $M_\odot$ pc$^{-2}$, for galactocentric radii between 4 and 13 kpc. It decreases as $1/R$ further away. Figures 3.12 and 3.13 show how its thickness varies in the Galaxy.

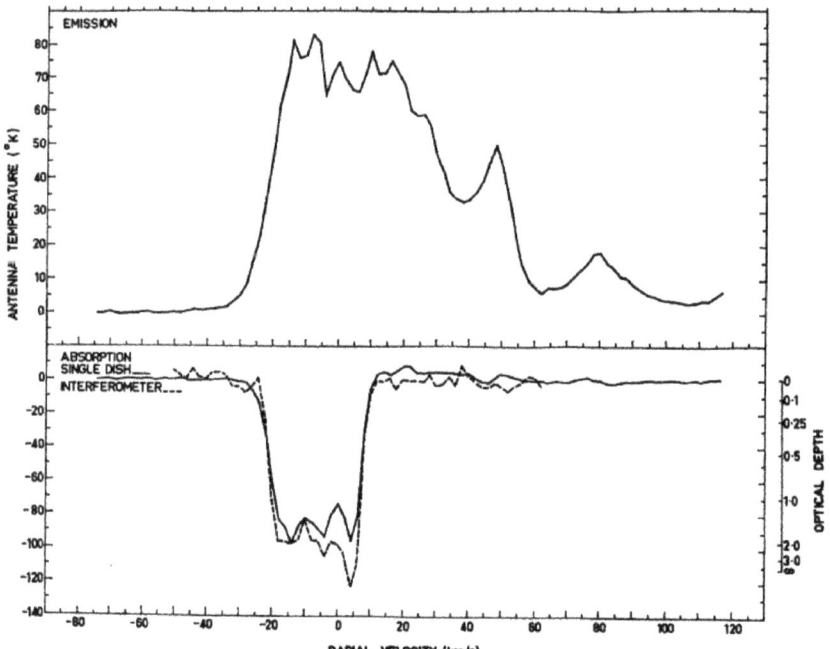

FIG. 3.10 – Profile of the 21-cm line in absorption in front of the radio source RCW 49 (l = 284.3°, b = -0.3°) and in emission at the immediate vicinity. The absorption (in% of the continuum flux of the radio source) is measured in two ways: with the 64-m radio telescope of Parkes in Australia (full line), and with an interferometer (dashed line). There is no absorption for radial velocities exceeding 5 km/s, which yields an approximate distance to the radio source, of 5 kpc. The absorption is very saturated, and the brightness temperature of the measured emission is then equal to the kinetic temperature of the neutral hydrogen, about 70 K. From Goss, W.M., Radhakrishnan, V., Brooks, J.W. & Murray, J.D. (1972) *Astrophysical Journal Supplement*, 24, 123-159, with permission of the American Astronomical Society.

# Structure and components of the Milky Way

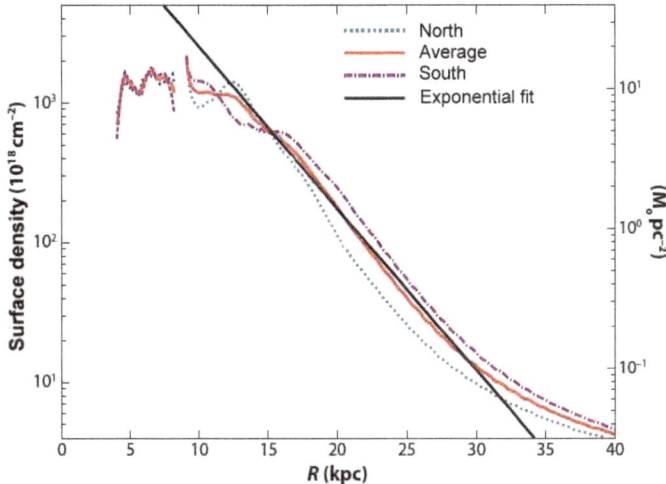

FIG. 3.11 – Radial distribution of the atomic component of the interstellar gas in the Galaxy. The projected surface density on the galactic plane of the number of hydrogen atoms (left) and the corresponding mass projected density (right) are plotted on a logarithmic scale as a function of galactocentric radius. An exponential does not fit the data well: the distribution of HI gas rather decreases as $1/R$. The Sun is assumed to be at a radius of 8.5 kpc. From Kalberla, P.M.W. & Dedes, L. (2008) *Astronomy & Astrophysics* 487, 951-963, with permission of ESO.

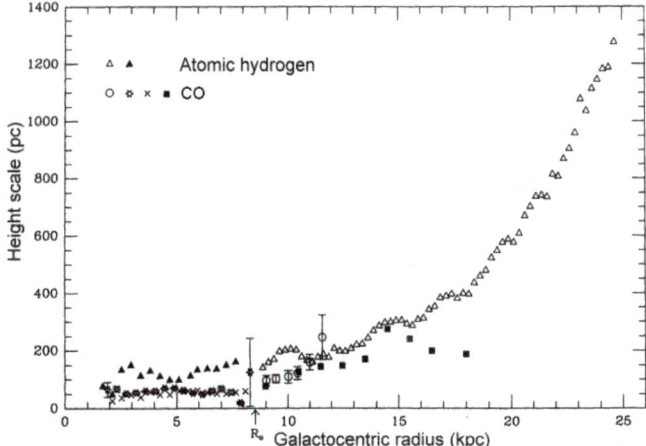

FIG. 3.12 – Radial variation of the thickness of the atomic and molecular gas in the Galaxy. The height scale is defined as the half thickness at a density equal to half the maximum density. The average values are given here only for a part of the Milky Way (longitudes 80°-280°). For the flaring of the disk, see Figure 5. From Wouterloot, JGA, Brand, J., Burton, WB & Kwee, KK (1990), *Astronomy & Astrophysics* 230, 21-36, with permission of ESO.

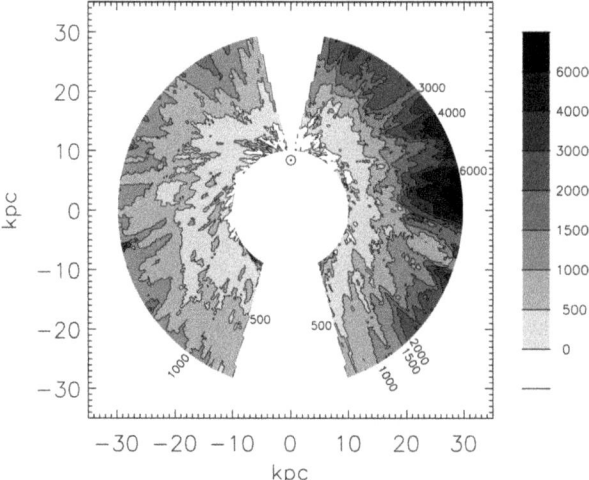

FIG. 3.13 – Map of the thickness of the atomic gas in the exterior of the Galaxy. It is asymmetric, clearly related to the warping of the disc (Fig. 3.5), and somewhat corrugated. From Levine, G.S., Blitz, L. & Heiles, C. (2006) *Astrophysical Journal* 643, 881-896, with permission of the American Astronomical Society.

Interstellar absorption lines offer another way to study the atomic component of interstellar gas: we saw an example at the end of the previous chapter. They allow in particular determination of its degree of ionization by measuring the abundance ratio of ion/neutral pairs as $Ca^+/Ca$, and also to obtain the chemical composition of the gas. Figure 3.14 compares the abundance of various elements in the line of sight to the star ξ Persei with respect to their abundance in the Sun. We can see that most elements are under-abundant in the interstellar medium, sometimes by very important factors, and more so if their temperature of condensation is higher. These elements are therefore condensed as dust either in envelopes of cool stars, or in the interstellar medium itself. The under-abundance is less marked in the warm neutral medium, indicating evaporation of grains, probably under the influence of shocks, which return some of their elements in the gas phase.

### 3.3.2 The molecular medium and the interstellar dust

We now know more than 150 molecules in the interstellar medium, mainly discovered through their rotational lines in millimeter and submillimeter waves. They are in cold (10-30 K) and relatively dense (more than 1000 molecules/$cm^3$) "clouds", which have a very complex, probably fractal structure. These fragments are often grouped in large complexes called giant molecular

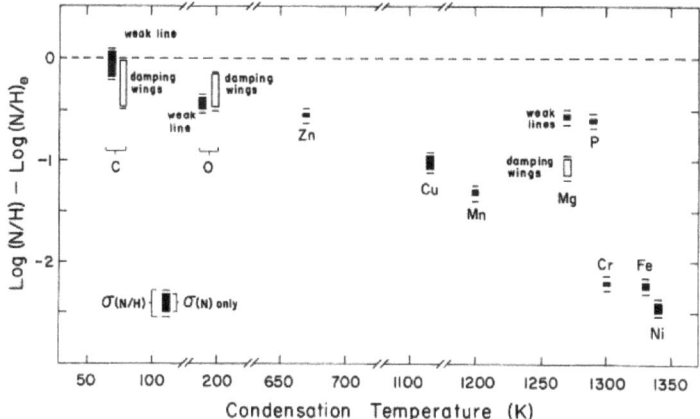

FIG. 3.14 – Abundance of certain elements in the neutral interstellar medium in front of the star ξ Persei, measured from the ultraviolet spectrum obtained with the Hubble Space Telescope. This is given in logarithmic units with respect to the solar abundance. The vertical dimensions of the symbols indicate the error. Only unsaturated weak absorption lines have been used, but for C, O and Mg one also indicates the abundance deduced from the damping wings of very saturated lines. For Mg, the difference probably results from an error in an atomic parameter. The condensation temperature of the element is plotted on abscissa. From Cardelli, J.A. et al. (1991) *Astrophysical Journal* 377, L57-60, with permission of the American Astronomical Society.

clouds. Some molecules are also present in the atomic neutral medium. The molecules are formed either by a complex chemistry, mainly ionic, or on dust grains that act as catalysts (this is particularly true of $H_2$). The dust grains are also covered with ices of water, methane, ammonia, methyl alcohol, etc., when they are buried deep in molecular clouds. The molecules are heated by various processes, mainly by the collision of protons of cosmic radiation with an energy of several tens of MeV, and cooled by line emission or by heat exchange with the grains.

The most abundant interstellar molecule is certainly $H_2$. Unfortunately, it is not directly observable, because it's rotational and vibrational transitions are strongly forbidden and the energy of the first rotational level is high: it can only be directly observed in hot gas, i.e. in shocks. For information on the molecular medium, and to obtain its mass, we have to use tracer molecules, the most useful by far beingt carbon monoxide (CO) molecule, whose first rotational lines at 2.6 and 1.3 mm are observed extensively. The problem is to obtain the $H_2$ column density of the gas from the intensity of these lines, which are usually highly saturated. We are led to also observe the less saturated lines of the isotopically substituted molecules

$^{13}C^{16}O$, $^{12}C^{18}O$: $^{13}C^{16}O$ is about 76 times less abundant than $^{12}C^{16}O$ and $^{12}C^{18}O$ about 560 times less abundant. It is impossible here to dwell on the various ways to address this problem, the solution remains tainted by significant uncertainties, and we give only the most recent result:

$$X = N(H_2)/W_{CO} \approx 2.54 \times 10^{20} \text{ mol.cm}^{-2} \text{ (K km/s)}^{-1}, \qquad (3.9)$$

where $N(H_2)$ is the column density of $H_2$ and $N(H_2)/W_{CO} = \int T^* dV$ the integrated intensity of the line of CO at 2.6 mm (115 GHz).

We can also attempt to use the gamma-ray emission from the interstellar medium. The interaction between cosmic energy protons of a few hundred MeV and interstellar nuclei (whether in the form of atoms, molecules or dust) produces gamma rays. Another source of gamma rays, which is not easy to separate from the previous one, is the decelerating radiation (*Bremsstrahlung* from German) of cosmic electrons in the electrostatic field of nuclei. The gamma-ray emission, which has been mapped by different satellites (see Fig. 1.8) has an intensity proportional to the density of cosmic particles and to the column density of interstellar matter in all forms. In principle, we can deduce the distribution of interstellar matter from a celestial map of the gamma rays, and by subtracting the contribution of atomic gas, derive that of the molecular gas (the contribution of dust is negligible). However, to what extent the density of protons and cosmic electrons is constant in the Galaxy in unclear, and there are discrete sources of gamma radiation that add to the contribution of interstellar matter, so this method is not very safe. It nevertheless contributed to the determination of the parameter $X$ of equation (3.9). Moreover, its application to small areas where we can assume that the density of cosmic particles is uniform revealed the existence of a "missing gas" in addition to that which is found from the 21-cm and CO lines. Finally, the gamma emission occurs in the outside of the Galaxy as a ring, the origin of which will be discussed in Section 3.6.

Another , more secure, method to determine the amount of interstellar material in the Galaxy is to study the distribution of interstellar dust and infer that of the gas assuming a constant dust/gas ratio, which seems to be justified (in other words, dust and gas are well mixed). This method has for several decades used the extinction, or rather the reddening of starlight by dust, measured by observing the stars at different wavelengths from the visible to the infrared, to map the column density in a molecular cloud: it is then assumed that the stars whose radiation is reddened are all located behind the cloud (see example Fig. A1.4). This method works well as long as one does not reach very deep areas of the cloud (column density $N(H) = 2N(H_2) > 10^{23}$ atoms cm$^{-2}$): then the dust grains are covered with ice, become larger, and they have optical properties different from those in the outer regions.

More recently, one has been able to use the thermal emission of dust grains, which are heated by stellar radiation and re-emit the absorbed energy at submillimeter wavelengths. It is then necessary to determine the temperature of the dust and its submillimeter optical properties to derive its column density, hence that of the gas. This has been made possible thanks to the Planck and COBE satellites that provided complete maps of the sky at various millimeter and submillimeter wavelengths. Knowing the optical properties of dust at these wavelengths by observing reference regions, the temperature $T_D$ of the dust is obtained at each point in the map from the ratios of intensities at different wavelengths. Then the optical thickness $\tau_D(\nu)$ (almost always small compared to 1) of its emission at frequency $\nu$ is obtained by the relationship:

$$\tau_D(\nu) = I\nu / B\nu(T_D), \qquad (3.10)$$

where $I\nu$ is the observed intensity and $B\nu(T_D)$ the brightness of the blackbody at frequency $\nu$ and temperature $T_D$.

The resulting maps of dust optical thickness at the peak of its thermal emission (545 GHz) and of dust temperature are presented on Figure 3.15, together with maps of the 21-cm and CO(1-0) line intensity. The observation of reference areas where we can be certain that there is only atomic hydrogen provides the optical thickness of the dust emission per atom $[\tau_D(\nu)/N(H)]_{ref}$. Then this optical thickness is correlated point by point in the map with that of the atomic + molecular gas: the atomic gas column density is derived from the intensity of the 21-cm line (equation (3.8)) and that of the molecular gas from that of the line of CO at 2.6 mm (equation (3.9)). This correlation is limited to galactic latitudes $|b| > 10°$, thus to regions of the solar neighborhood where the 21-cm and CO lines are optically thin. So we have:

$$\begin{aligned}\tau_D(\nu) &= [\tau_D(\nu)/N(H)]_{ref} [N(H) + 2N(H_2)] \\ &= [\tau_D(\nu)/N(H)]_{ref} [N(H) + 2XW(CO)].\end{aligned} \qquad (3.11)$$

The result is shown in Figure 3.16. There is an excellent correlation between $\tau_D(\nu)$ and the gas column density $N(H) < 8 \times 10^{20}$ atom cm$^{-2}$, corresponding to regions dominated by the atomic gas. The correlation is also very good for areas where the gas column density $N(H) = 2N(H_2) > 5 \times 10^{21}$ atom cm$^{-2}$, in all directions completely dominated by the molecular gas; but it degrades for very high column densities, where the CO line can no longer trace any gas due to its huge optical thickness, and where the properties of dust begin to change. The correlation is the same in both regimes if we take $X = (2.54 \pm 0.13) \times 10^{20}$ mol.cm$^{-2}$ (K km/s)$^{-1}$. This is the best value available for this parameter in the solar neighborhood, which is reported in equation (3.9).

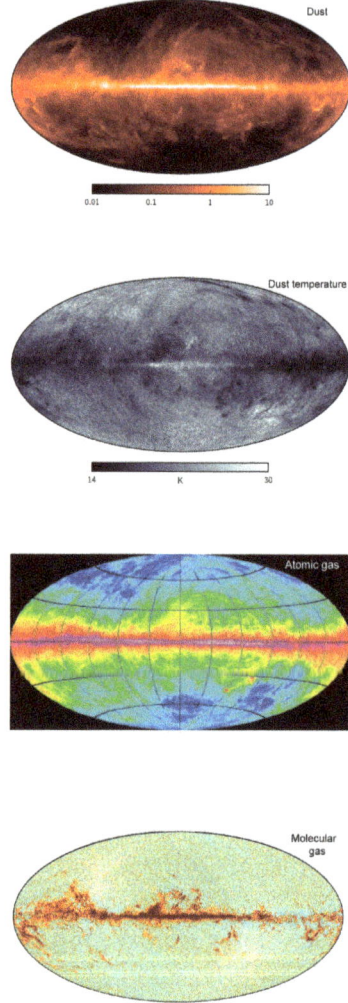

FIG. 3.15 – From top to bottom, full-sky maps of the column density of dust (given as the optical depth at 345 GHz in a logarithmic scale), of the temperature of dust (linear scale), of the column density of atomic hydrogen (from the 21-cm line intensity, logarithmic scale), and of molecular gas (from the CO(1-0) line intensity, logarithmic scale). The Galactic center is at the middle, and the galactic longitude increases from left to right from -180° to +180°. The 21-cm line and the CO(1-0) line being saturated at low galactic latitudes, the gas column densities are underestimated there. Conversely, the column density of dust is correctly obtained everywhere. From Kalberla, P.M.W. et al. (2005) *Astronomy & Astrophysics* 440, 775-782 (atomic gas) and from the Planck collaboration (2015) *Astronomy & Astrophysics*, in press (2015arXiv150201588P, other data), with permission of ESO.

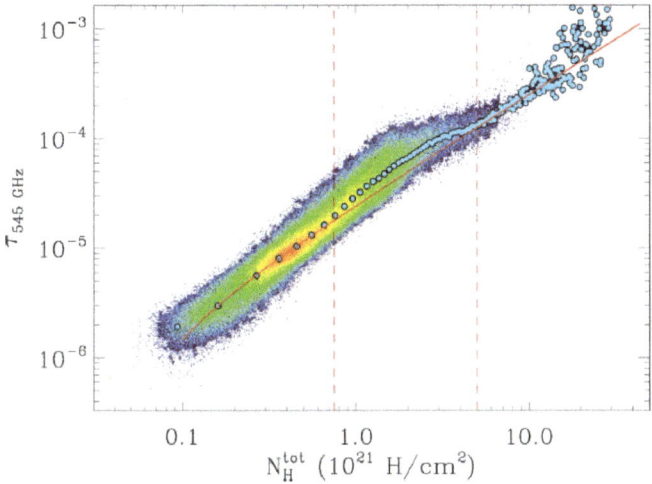

FIG. 3.16 – Correlation between the optical thickness of the interstellar dust at the frequency of 545 GHz (wavelength 0.55 mm) and the total gas column density. The small blue circles indicate average values for different gas column densities. There is a good correlation for the low values of this column density and also for high values (red line), and an excess of optical thickness for the intermediate values. From the Planck Collaboration (2011) *Astronomy & Astrophysics*, 536, A19, with permission of ESO.

The intermediate column density region corresponds to an excess of dust opacity, corresponding to the "missing gas" mentioned previously, which was already suspected from observations of gamma rays. There is a simple explanation of this phenomenon: when ultraviolet radiation from stars enters a molecular gas cloud, there is a region where the molecular hydrogen is not completely photo-dissociated while CO is fully dissociated into $C^+$, C and O. The missing gas is molecular hydrogen without CO. In the solar neighborhood where the observations are done, its mass is 28% of that of the atomic gas and 128% of that of the molecular gas detected by the CO line, which is relatively low. It is not yet possible to know how much of this missing gas the Galaxy contains in total, but it is clear that the estimates of the amount of molecular gas in the galaxy made from CO lines are significantly undervalued, probably by at least 30%. The total mass of molecular medium should be in the order of $3 \times 10^9$ $M_\odot$, about as much as the cold atomic gas, but much more might exist in the external regions of the Galaxy as will be discussed later in Section 3.6.2. The radial distribution of the surface density of the molecular medium in the Milky Way is shown Figure 3.17. This density decreases much faster than that of the atomic medium (Fig. 3.11), but low CO emissions were still detected in the outermost regions of the Galaxy, to a galactocentric radius of 21 kpc. The thickness of the molecular gas in the disk is lower than that of the atomic gas: see Figure 3.12.

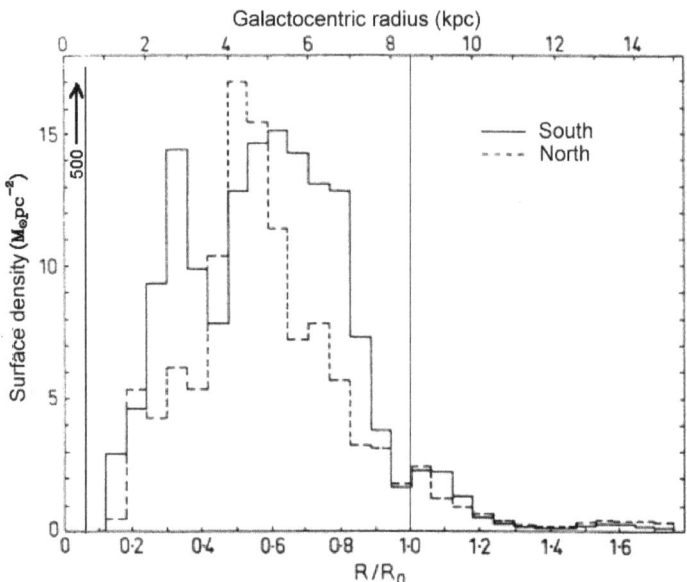

FIG. 3.17 – Radial surface distribution of the molecular medium. It is obtained from the intensity of the CO line at 2.6 mm by Robinson et al. (1988), *Astronomy & Astrophysics* 193, 60-68 (with permission of ESO), taking $X = 2.5 \times 10^{20}$ mol cm$^{-2}$ (K km/s)$^{-1}$. 30% was added to the mass thus calculated to take into account the missing molecular gas. The surface density is very high near the galactic center where the molecular gas mass is about $4 \times 10^8$ $M_\odot$ within less than 500 pc. The surface density is certainly underestimated beyond the solar radius, because the C/H and O/H abundance ratios decrease outward, the CO/H$_2$, and there may be a very large quantity of molecular hydrogen without CO.

### 3.3.3 The ionized medium

Part of the interstellar hydrogen is ionized by the ultraviolet radiation of the hot and massive stars of type O and B. These are formed in molecular clouds and have a limited life time, of a few million years (see Tables 2.2 and 2.3), so they stay close to the remnant of their parent cloud. They ionize the nearby medium, which can be relatively dense. The photons that can ionize hydrogen, whose wavelength is shorter than 91.2 nm (energy larger than 13.6 eV), are absorbed very easily by the hydrogen atoms, whose effective ionization section is very large: the small amount of neutral hydrogen that is formed by the recombination of protons and electrons in the ionized medium is immediately re-ionized, and this occurs up to the star distance where no more ionizing photons remain. There is therefore a more or less spherical region (the *Strömgren sphere*) around the ionizing star, or the ionizing star cluster, where hydrogen is almost completely ionized, while remaining neutral on the outside: this is a gaseous nebula, also called HII region.

HII regions emit hydrogen recombination lines, including those of the Balmer series that fall into the visible range. Other recombination lines correspond to transitions between very high levels and fall in the radio range. These lines are very useful for obtaining the radial velocity of the nebula, while one can get the distance of the exciting stars by photometry: these parameters were used to build the map of the Galaxy of Figure 1.6. Furthermore, HII regions emit many fine-structure lines of different ions, which are excited by collision with the free electrons. They are forbidden by the selection rules and are generally unobservable in the laboratory: the probability of spontaneous emission from the upper level of the transition is very small, but as this level can only be populated by collisions, if the density is low, radiative transitions eventually occur. These lines are therefore very sensitive to the density that they serve to measure. They also provide an abundance of the ions that produce them, and help to infer the gas temperature (of the order of 10 000 K) from the abundance ratio of ion pairs like $O^+$ and $O^{++}$.

The abundances of He, C, N, O, Ne, and S, in HII regions close to the Sun, like the Orion Nebula, are very similar to those in the solar system. Only iron is deficient by an order of magnitude. This indicates that the dust grains, which contain a large proportion of C, N and O in the neutral medium, are destroyed in the HII regions, with the exception of the most refractory grains containing iron. The abundances in HII regions show a gradient with galactocentric radius: they decrease towards the external parts of the disk. This provides a valuable tool to study the evolution of the Galaxy. We will return to this topic in Chapter 7.

HII regions are often in contact with what remains of the molecular cloud after the exciting stars were formed. The transition zone is subjected to an intense ultraviolet radiation from these stars. The UV photons of energy less than 13.6 eV ionize some elements and dissociate molecules until the radiation is totally absorbed by these ionizations and dissociations, and by dust mixed with the gas. The structure of these regions, commonly referred to as *photodissociation regions* (PDR), is represented in Figure 3.18. Photodissociation regions are extremely bright in the $C^+$ lines at 158 mm, in the O lines at 63 and 145 mm, and also in the $H_2$ vibration lines in the near infrared; the dust is strongly heated so that its thermal radiation in the mid and far infrared is also very intense. Note that these photodissociation regions are not fundamentally different from the regions that we mentioned previously, where the "missing" molecular gas with no CO is hidden: they are simply much brighter in the infrared because of the particularly intense ultraviolet radiation, and their temperature is higher.

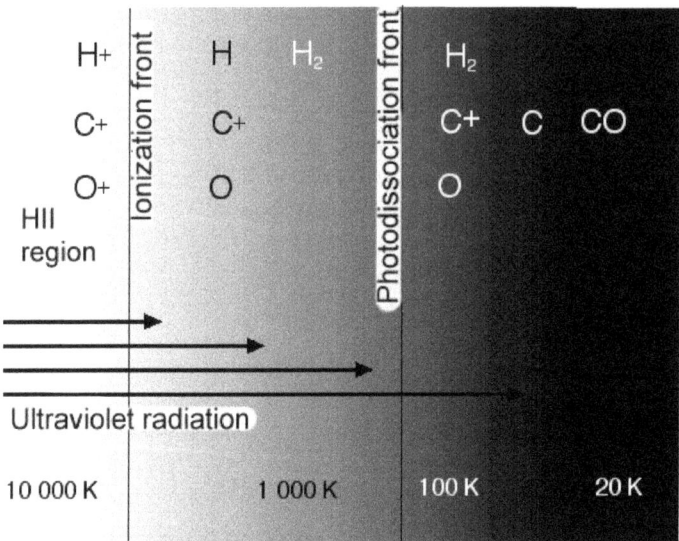

FIG. 3.18 – Structure of a photodissociation region. The ultraviolet radiation of the stars that ionized the HII region enters the adjacent molecular cloud, but it weakens progressively due to the absorption by the ionization of some elements and the dissociation of molecules it produces, and by the dust mixed with the gas. The HII region is limited by an ionization front, beyond which there are no more photons with energy higher than 13.6 eV. Hydrogen is thus atomic and oxygen is neutral, while carbon and metals are ionized. Molecular hydrogen, highly resistant to photodissociation, survives to some depth. Deeper, beyond the photodissociation front, there are still other molecules, but CO is partly photodissociated into C, $C^+$ and O. Deeper, it is not affected. The typical gas temperature of these various regions is indicated.

The pressure in the HII regions is much greater than outside, because of their high temperature. Therefore, a HII region is always expanding, and eventually leaks into the interstellar medium from the location of its periphery where it meets the least resistance (*champagne effect*). It is this phenomenon that limits the lifetime of HII regions to about $10^4$ years on average, a time much shorter than the lifetime of the exciting stars. The released gas stays ionized for a long time because the characteristic time of the hydrogen recombination is of the order of $6 \times 10^6$ years at the atom density, 0.03 $cm^{-3}$, encountered in the interstellar medium outside clouds. Oddly enough, the gas temperature remains high despite its adiabatic expansion, about 8000 K, which indicates the existence of a still unknown heating source, perhaps the dissipation of plasma turbulence. The total mass of the diffuse ionized gas in the Galaxy is quite uncertain, probably on the order of $5 \times 10^8$ $M_\odot$.

Figure 3.19 illustrates the distribution of the thermal emission of the ionized gas in the whole sky. It is proportional to the *emission measure* $\int n_e^2 dL$ over the line of sight, expressed in cm$^{-6}$ pc.

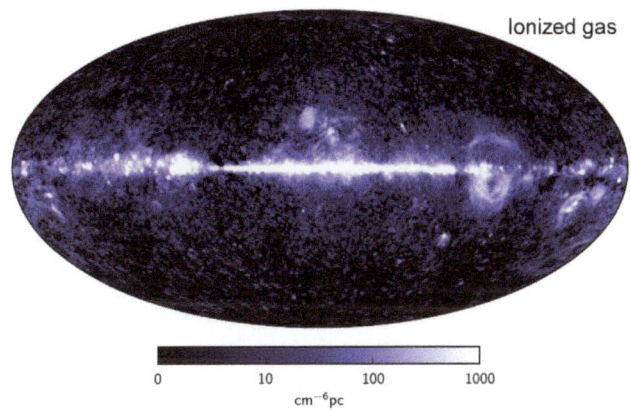

FIG. 3.19 – Full-sky map of the thermal emission of the ionized gas (presentation as for Fig. 3.15). The emission measure $\int n_e^2 dL$ over the line of sight, expressed in cm$^{-6}$ pc, is plotted in a logarithmic scale from radio continuum observations. The strong emission of HII regions is conspicuous, but we can also see the weak emission of the low-density, diffuse ionized gas. From the Planck collaboration (2015) *Astronomy & Astrophysics*, in press (2015arXiv150201588P), with permission of ESO.

### 3.3.4 Supernova remnants, bubbles and hot gas

Massive stars end their lives in a violent explosion as supernovae. Another type of supernovae (SN Ia) is formed by the explosion of the accreted material on a white dwarf star belonging to a tight binary system. Supernovae occur on average every 50 years in a galaxy like ours. In both types, the kinetic energy of the ejected envelope is of the order of $10^{50}$ ergs, the largest part of which is used to stir the interstellar medium. This envelope expands rapidly, at a velocity that can reach 20,000 km/s at the beginning of the expansion, and is limited by a shock. It heats the interstellar medium it crosses, so that within the envelope of the supernova remnant there is gas at more than $10^6$ K. In addition, the shock, which contains a magnetic field originating from the star or from the interstellar medium, accelerates charged particles, ions and electrons, to enormous energies: this is the origin of most cosmic particles.

After about a million years, the supernova remnant is dispersed into the surrounding environment. The internal gas has somewhat cooled by adiabatic expansion, but its temperature remains between $10^5$ and $10^6$ K and is fully ionized; heavy ions are also multi-ionized. We find this very hot gas in the interstellar medium, where it occupies a significant fraction estimated at about 30% by volume in the galactic plane, with a density of about $10^{-3}$ ion cm$^{-3}$: it emits X-rays, and its highly-ionized ions (NV, OVI) produce absorption lines in the far ultraviolet. Despite the low density, its pressure is high so it spreads out in the halo of the Galaxy; it cools after some $10^8$ years, condenses and falls back to the galactic plane (galactic fountain).

As massive stars are rarely isolated but are generally grouped into associations, several stars of an association may explode in turn as supernovae. In addition, O stars lose a lot of material before exploding, as a hot stellar wind ($10^5$-$10^6$ K) whose expansion velocity is of the order of 2000 km/s and the total energy is comparable to that of a supernova. The collective effect of these explosions and winds form a giant bubble (100-1000 pc diameter), expanding at a rate of several tens of km/s. Its total kinetic energy can exceed $10^{53}$ ergs, much more than that of an isolated supernova. These bubbles are filled with very hot gas, which radiates X-rays, while their expanding envelope is optically visible. We saw in the previous chapter that we are located inside a bubble, which is rather old because the stars that formed it have disappeared. After a few tens of millions of years, the bubbles eventually disperse into the general interstellar medium, such as single supernova remnants, and thus contribute to the hot gas that we just discussed. This dispersion often takes the form of a chimney perpendicular to the galactic plane, and at some places the gas is ejected from the Galaxy into the intergalactic space.

## 3.4 Radiation fields, magnetic field, cosmic particles and radio radiation

On the Earth we are bathed by electromagnetic radiation at all wavelengths. The radiation from the distant Universe is relatively weak, with of course the exception of the cosmic blackbody radiation at 2.726 K, which completely dominates at the submillimeter and millimeter wavelengths from 500 mm to 2 mm. Figure 3.20 shows schematically the intensity of electromagnetic radiation at different wavelengths, as observed at high galactic latitudes, disregarding what we receive on the Earth from the upper atmosphere and the solar system (zodiacal light). So, it comes from the stars and the interstellar medium. This Figure does not show the weak emission at centimeter to decameter radio waves and the X-ray and gamma emissions, which we will discuss later. It is quite difficult to determine the total amount of electromagnetic radiation emitted by the Galaxy, given our disadvantaged position, but we estimate that half of the stellar radiation is absorbed by dust and re-radiated in the far infrared. The ambient ultraviolet radiation, which plays a big role in the physics of the interstellar medium, is displayed in Figure 3.21. It corresponds to what we see in the solar neighborhood, but it can obviously be very different in other parts of the Galaxy, in particular close to HII regions. We will only mention in passing the X radiation that originates from discrete sources (stars and supernova remnants), from the hot gas, and from the relativistic cosmic electrons: at low (keV) energies, it is very variable from one place to another in the Galaxy, due to absorption by the interstellar medium; see Figure 1.8, where this absorption is visible at low galactic latitudes.

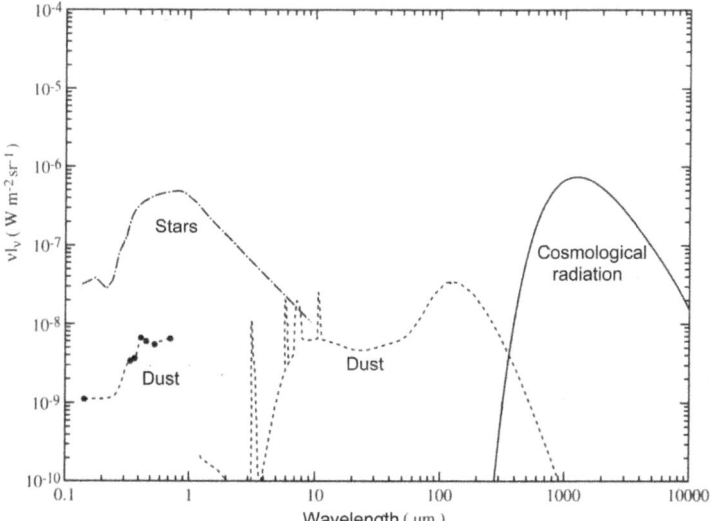

FIG. 3.20 – The electromagnetic radiation observed at high galactic latitudes, from ultraviolet to millimeter waves. The product $I_\nu$ of the monochromatic luminance of the sky by the frequency $\nu$ is plotted as a function of wavelength for various sources of radiation. The advantage of using this product $\nu I_\nu = \lambda I_\lambda$ is that this quantity considered in a logarithmic interval of frequency (or wavelength) directly represents the total energy in this interval. Dust diffuses the starlight in the visible (dots and dotted lines at left) and gives a thermal emission in the infrared. The bands at 3.3, 6.2, 7.7 and 11.3-12.7 mm and the underlying continuum are due to out-of-equilibrium thermal emission by very small hydrogenated polycyclic aromatic grains (PAHs)[4]. The continuum at longer wavelengths is the thermal emission of the larger dust grains. From Leinert, Ch. et al. (1998) *Astronomy & Astrophysics Supplement* 127, 1-99, with permission of ESO.

---

[4] The ionized PAHs in rotation also produce a relatively weak radio emission from 10 to 100 GHz that we only mention in passing here.

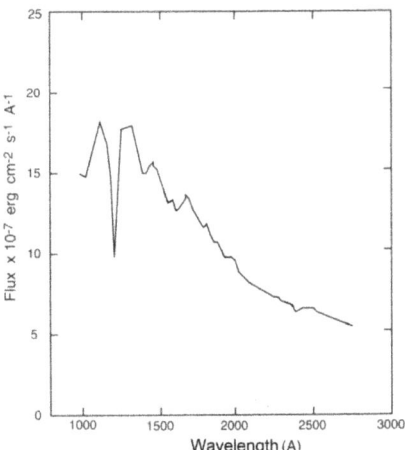

FIG. 3.21 – The ultraviolet radiation from the sky. The integrated flux over the entire celestial sphere around the Sun is plotted against the wavelength. It comes from the hot stars. This flux is essentially zero at wavelengths less than 912 Å (91.2 nm), due to absorption by atomic hydrogen. The minimum at 1216 Å corresponds to absorption in the Lyman α line of hydrogen. From Gondhalekar, P.M., Philips, A.P. & Wilson, R. (1980) *Astronomy & Astrophysics* 85, 272-280, with permission of ESO.

The interstellar medium is everywhere and is a conductor of electricity: there are always electrons in the free state, since metals and carbon are ionized by ultraviolet radiation in the HI medium. Even molecular clouds are conducting because cosmic rays create some ionization. The entire interstellar medium can contain, and actually contains, a magnetic field whose strength is about 3 microgauss ($3 \times 10^{-10}$ tesla) in the diffuse medium. This can be measured by the Faraday rotation of the polarization plane of radio waves in an ionized diffuse medium, or from the galactic synchrotron radiation (see below), or finally by the polarization of starlight that is produced by dust grains oriented by the magnetic field. We observe a large-scale component of about 1.4 microgauss oriented parallel to the regular spiral arms, superimposed onto a random component with a similar intensity, which reflects the turbulent motions in the interstellar medium. In the dense clouds, it is possible to measure the magnetic field by the Zeeman effect on the 21-cm line or on radio lines of the OH and CH molecules. We thus observe that there is no conservation of the magnetic flux when clouds collapse, indicating a partial removal of the magnetic field by the ambipolar diffusion which occurs inevitably in the plasma. The magnetic field plays a less and less important role when a cloud collapses to form stars.

Another component of the interstellar medium that we have only mentioned in passing is the cosmic radiation: although generally used, the term "radiation" is particularly inappropriate since it is made of charged particles, which are nuclei, electrons and positrons. All elements are found there, but with abundances somewhat different from abundances in the Sun or the interstellar medium: the heavy elements are generally more abundant relative to hydrogen, and the light elements Li, Be and B, which are easily destroyed by stars and therefore rare, are also abundant in cosmic rays. Figure 3.22 shows the energy spectrum of three nuclei (H, He, and Fe), as observed outside the Earth's atmosphere between 0.01 and 100 billion electron volts per nucleon (GeV/n). The time-variable drop of intensity at low energies is due to the modulation by the magnetic field associated with the solar wind; at energies below a few tens of million electron volts (MeV), the rise corresponds to cosmic rays of solar origin, and has nothing to do with the galactic cosmic rays. Therefore the flux of cosmic particles at energies below about 3 GeV/n is not directly known, and the Figure shows only a possible extrapolation at these energies. Between 3 and 300 GeV/n, the energy spectrum, which is well known, is represented by the power law:

$$I(E) = 1.34 \times 10^4 \, E^{-2.6} \, \text{m}^{-2} \, \text{s}^{-1} \, \text{sterad}^{-1} \, (\text{GeV}/\text{n})^{-1} \;, \tag{3.12}$$

$I(E)$ being the flux and $E$ the energy in GeV/n.

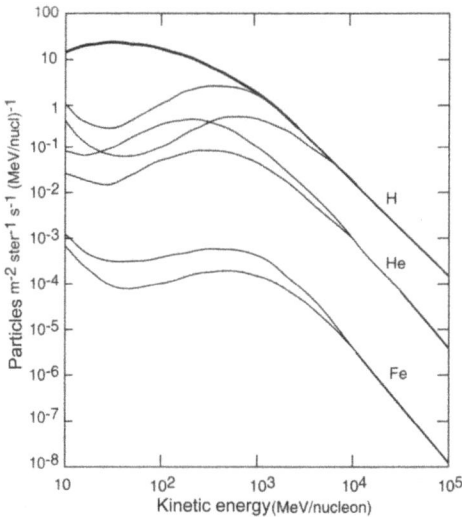

FIG. 3.22 – Energy spectrum of cosmic rays observed for protons (H), α particles (He) and iron nuclei. Below $10^4$ MeV/nucleon, the curves are split: the higher curve corresponds to the flux at the solar minimum and the lower one to that at the maximum of solar activity (solar modulation). At very low energies, the rise is due to cosmic rays of solar origin. The bold curve is an estimate of the spectrum of protons outside the solar cavity, due to Prantzos. From Silberberg, R. & Tsao, CH (1988) in *Genesis and Propagation of Cosmic Rays*, ed. Shapiro, M.M. & Wefel, J.P., Reidel, Dordrecht, p. 41.

All cosmic particles are accelerated in the magnetic shock surrounding supernova remnants, at least up to 300 GeV/n. In addition, there are cosmic particles with extremely high energies up to $10^{20}$ eV, whose origin is probably extragalactic.

It is interesting that there is an approximate equality in the plane of the Galaxy between the energy density of cosmic rays $u_C$, the density of the kinetic energy $u_{turb}$ of the interstellar gas that is dominated by macroscopic movements, mainly turbulent, and finally that of the magnetic field $u_B$. Each is about $4 \times 10^{-12}$ erg cm$^{-3}$, or 2.5 eV cm$^{-3}$. This is of course not a coincidence. The reason is that the cosmic rays excite in the interstellar medium, which is a plasma where the magneto-hydrodynamic approximation is applicable, Alfvén waves which in turn diffuse the gas particles.

In addition to the cosmic ray nuclei, we detect in the vicinity of the Earth electrons with energies between 1 and $10^3$ GeV or more. Their flux at 10 GeV is 1/100 of that of the protons. They too are accelerated in the shock of supernova remnants. About 1% of the light particles are positrons created by reactions between high energy cosmic protons and interstellar nuclei: these reactions produce, inter alia, $\pi^+$ and $\pi^-$ mesons, which decay into charged muons that themselves decay into positrons and electrons, respectively. The energy spectrum of electrons, as observed in the solar environment, which is therefore affected by solar modulation and especially by synchrotron losses in the magnetic field carried by the solar wind, is represented by a power law:

$$I(E) = 700 \, E^{-3.3} \, \text{m}^{-2}\,\text{s}^{-1}\,\text{sterad}^{-1}\,\text{GeV}^{-1} \qquad (3.13)$$

Being charged, the cosmic particles follow trajectories determined by the galactic magnetic field: they spiral around the lines of force with a gyration radius $r$ given by:

$$r = 3.33 \times 10^{12} \, A/Q \, E/B \text{ cm}, \qquad (3.14)$$

for a relativistic particle. $A$ is the mass number, $Q$ the charge number, $E$ the energy per nucleon in GeV/n and $B$ the magnetic field in mG. For example, a $10^6$ GeV proton in a 5 mG field has a gyration radius of $10^{17}$ cm (0.2 pc). As a consequence, the cosmic particles are well confined in the Galaxy by the magnetic field and take a long time to escape: the study of their elemental composition and of the lifetime of radioactive isotopes as $^{10}$Be, which are formed by decomposition of heavier cosmic nuclei during their encounter with interstellar nuclei (*spallation* reactions), shows that they spend an average of twenty million years in the Galaxy before escaping, and that the column density along their trajectory corresponds to an encountered mass (the *grammage*) of approximately 7 g cm$^{-2}$. We can easily deduce from these figures that the cosmic particles stay for 1/5 of their life in the galactic disk, and spend the remaining time, before escaping, in the halo around the disc, which contains very little interstellar material.

The interaction of cosmic particles with the interstellar medium produces gamma rays by three different mechanisms. 1: nuclear interactions between high-energy nuclei and interstellar nuclei, which produce among other neutral $\pi^0$ mesons, decaying into two gamma rays; 2: deceleration radiation (*Bremsstrahlung*) of cosmic electrons in the electrostatic field of nuclei; 3: inverse Compton effect, that is to say, inelastic scattering of high-energy electrons by low-energy photons, essentially those of the cosmic radiation background except near intense sources (see Fig. 3.20). Figure 3.23 illustrates these respective contributions as a function of particle energy. We have seen that the gamma-rays observation has allowed, to some extent, the determination of the mass of the interstellar medium and its large-scale distribution.

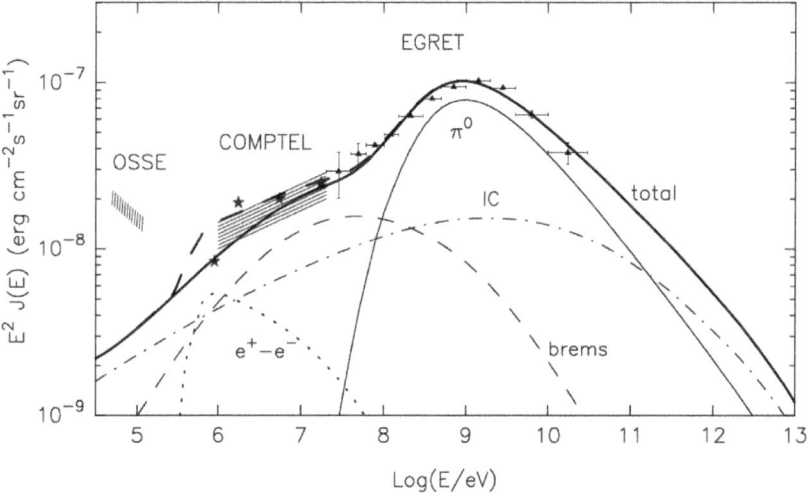

FIG. 3.23 – Energy spectrum of the diffuse emission of the inner regions of the galaxy in hard X-rays and gamma rays, observed with different satellites. To make the Figure more readable, the flux was multiplied by the square of the energy per nucleon, $E^2$. The observed spectrum is compared with the theoretical spectrum calculated from the energy spectrum of cosmic protons and electrons. The contribution of the three mechanisms mentioned in the text ($\pi^0 = \pi^0$ decay, brems = Bremsstrahlung, IC = inverse Compton) is indicated. The in-flight annihilation of electrons and positrons ($e^+ - e^-$) might give the possible excess between 1 and 10 MeV. The excess of hard X rays observed by OSSE is due to discrete sources of synchrotron radiation produced by high-energy electrons accelerated in supernova remnants and pulsars. These electrons also produce, by the inverse Compton mechanism, the very high energy gamma rays currently studied by the HESS instrument. Towards the outside of the Galaxy, the observed spectrum is different, which may indicate an electron deficiency, a softer spectrum of protons, or another production mechanism (see Section 3.6 of this chapter). From Aharonian, FA & Atoyan, AM (2000), *Astronomy & Astrophysics* 362, 937-952, with permission of ESO.

The interaction of cosmic electrons of relativistic energy ($E \gg 1$ MeV) with the galactic magnetic field produces radio photons by the synchrotron mechanism (magnetic Bremsstrahlung). The discovery in the early 1950s of this radiation, which occurs in numerous astrophysical situations, especially in supernova remnants, was, with that of the 21 cm line, the first major success of radio astronomy. The synchrotron emission is strongly linearly polarized, and it is this relatively rare property in astrophysics that allows it to be recognized without difficulty.

The characteristic frequency $\nu_c$ of the synchrotron radiation by electrons of energy $E$ in the magnetic field $B$ is:

$$\nu_c = 16.1 \ E^2 B \text{ MHz}, \qquad (3.15)$$

where $E$ is in GeV and $B$ in mG. Thus, a 2 GeV electron emits around 300 MHz in the typical galactic magnetic field of 5 mG. The energy spectrum $E$ of electrons, in power law,

$$\mathrm{d}n(E) = K \ E^{-\gamma} \ \mathrm{d}E \qquad (3.16)$$

yields a radio spectrum also in power law, as observed:

$$I(\nu) \propto K \ B^{(\gamma+1)/2} \ \nu^{-(\gamma-1)/2}. \qquad (3.17)$$

The synchrotron radio continuum spectrum of the Galaxy indeed obeys a power law with a slope of -0.75 which corresponds to $\gamma = 2.5$. Thus the slope of the energy spectrum of the relativistic electrons is similar to that of cosmic nuclei, with $\gamma = 2.6$ (see equation (3.12)), but differs from that observed for electrons in the vicinity of Earth, with $\gamma = 3.3$ (equation (3.13)): these electrons are in fact affected by synchrotron losses in the magnetic field of the solar wind. This warns us against the temptation to combine the flow of electrons observed near the Earth with the galactic synchrotron continuum to determine the magnetic field.

The continuum radio emission of the Galaxy is dominated at wavelengths longer than about 20 cm (frequency less than 1.5 GHz) by the diffuse synchrotron radiation, plus that of supernova remnants (Fig. 3.24); at shorter wavelengths, the free-free thermal radiation of the ionized gas dominates, while the thermal radiation of the dust takes over below 3 mm (100 GHz). The synchrotron radio radiation is much extended in galactic latitude compared to these thermal radiations, which means that it occupies a thick halo. It is difficult to study this halo in our Galaxy because of our unfavorable position and of the lack of distance criteria for radio waves, but we see it very frequently in spiral galaxies seen edge-on: an example is shown in Figure 3.25.

Structure and components of the Milky Way 71

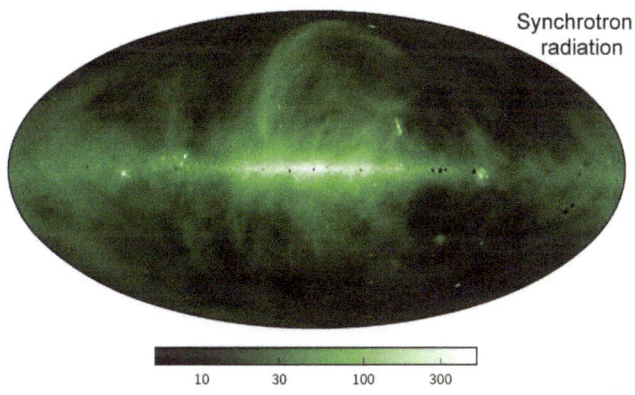

FIG. 3.24 – Full-sky map of the synchrotron emission at 408 MHz (presentation as for Fig. 3.15 and 3.19). Logarithmic scale of intensity (see the reference for the definition of the unit). Some supernova remnants are seen. The strongest extragalactic sources have been suppressed. Note the North Polar Spur, a very extended arc in the northern hemisphere, also visible in Figure 3.15: this is a giant interstellar bubble. Compare to the thermal radio radiation of Figure 3.19. From the Planck collaboration (2015) *Astronomy & Astrophysics*, in press (2015arXiv150201588P), with permission of ESO.

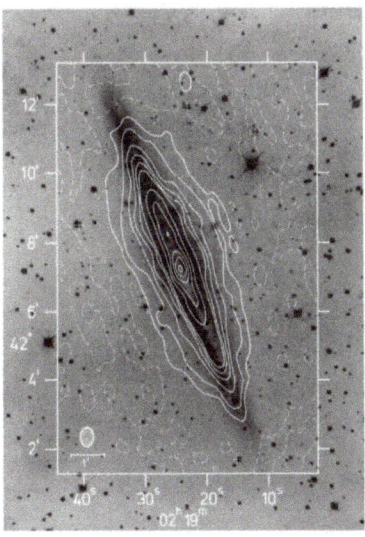

FIG. 3.25 – The radio emission of the edge-on spiral galaxy NGC 891. The white contours show the emission in the radio continuum 1412 MHz (21.2 cm), superimposed on an optical image. The small ellipse at the bottom left represents the angular resolution of the instrument (the Westerbork interferometer in the Netherlands). The emission is much thicker than the visible galaxy, forming a radio halo. From Allen, R.J., Baldwin, J.E. & Sancisi, R. (1978) *Astronomy & Astrophysics* 62, 397-409, with the permission of ESO.

## 3.5 The spiral structure of the Galaxy

A characteristic feature of the type of galaxies like our own is the existence of a spiral structure (Fig. 3.26). The presence of such a structure in a large number of galaxies raises a problem that can be summarized as follows: if a spiral structure exists at a given moment, it is destroyed after about $10^8$ years by the differential rotation. It is therefore necessary that some mechanism compensates for the effect of the differential rotation by stiffening the arms, or to admit that the spiral arms are compression zones of a wave that is stationary or quasi-stationary in a rotating reference system.

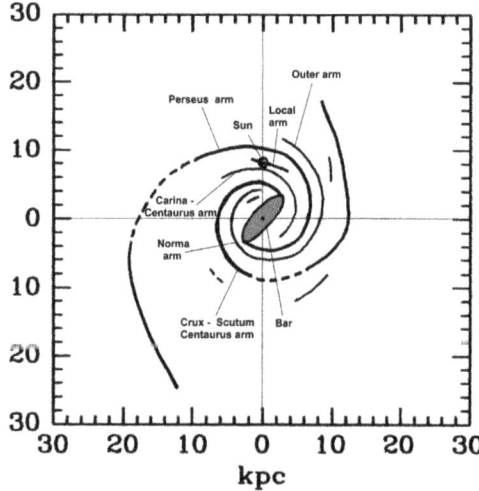

FIG. 3.26 – Diagram of the spiral structure of the Galaxy. The actual width of the arms is larger than that seen in this diagram, which only indicates their position.

Different theories have been devised in the past to "stiffen" the spiral arms. In particular it was assumed that the magnetic field governs the gas motion, the arms being magnetic force lines. In such a theory, the gas channeled by the magnetic field has to flow along the arms to compensate for the differential rotation. Then, there must be a neutral point where the gas is at rest relative to the arm: at this radius, there is co-rotation between the arm and the material of the disk as a whole. At larger galactocentric radii, the gas should flow outwards, and conversely flow inwards at smaller radii, which implies a continuous supply of gas at the neutral point. No such thing is seen, and moreover the predicted motions of the gas along the arms are not observed. Such theories, which had their heyday in the early 1960s, should be abandoned. Moreover, the magnetic field is better known than at that time, and is too weak to sustain a spiral structure.

We must therefore consider that the origin of the arms is gravitational. After the pioneering work by Bertil Lindblad (1895-1965) then by Alar Toomre in the early 1960s, the final solution came from the Sino-US astronomers Chia-Chiao Lin and Frank Shu in 1964: the spiral arms are density waves formed and maintained for billions of years by gravitational instability, or under the effect of gravitational perturbations that do not have cylindrical symmetry. In our Galaxy, the most significant such perturbation comes from the central bar. We will return to this in Chapter 5. For now, we will just mention that the velocity dispersion of the disk components stabilizes the matter against instabilities: we would thus obtain a sharp spiral structure in systems where the velocity dispersion is not too high, mainly that of gas and relatively young stars. Numerical simulations confirm these ideas.

We have seen that it is possible to trace to some extent the spiral arms from the radial velocity of the gas and HII regions, even better when in addition the distances of the emitting areas are known. However, to understand what is happening, it is better to turn to external galaxies because our position is very unfavorable. For example, anomalies of the gas velocity at the passage of the spiral arms are observed in the 21-cm line, implying that it does not follow a uniform rotation (Fig. 3.27). But the best confirmation of the idea that the arms are density waves comes from numerical simulations, which account very well for the observations.

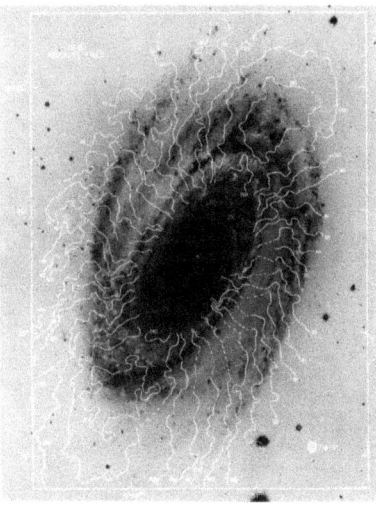

FIG. 3.27 – The radial velocity field of the galaxy M 81, superimposed on a visible light image. The small ellipse at the lower right shows the resolving power of the 21-cm line observations that provide the radial velocities. The isovelocity lines are often deflected in the arms, which indicates systematic motions relative to the rotation, caused by the passage of the spiral wave. From Rots, A.H. (1975) *Astronomy & Astrophysics* 45, 43-55, with permission of ESO.

Let us return to our Galaxy. Until recently, there was little information on the spiral structure of the stellar component of the galactic disk, whose existence was suspected because the gravitational waves affect the stars as well as the interstellar matter. The stellar disc is only seen clearly in the infrared, where the emission is not dominated by young stars and gaseous nebulae as it is in the visible, and where interstellar extinction is much smaller. The infrared images of nearby galaxies show a spiral structure in the stellar disk, but much simpler than the one we see in visible light (Fig. 3.28). Some secondary spiral structures hardly appear in the infrared.

FIG. 3.28 – The galaxy M 83 in the visible (left) and in the infrared (right). It is seen that only the two most pronounced spiral arms in the visible appear well in the infrared, the other structures being barely seen. The bar is also much more conspicuous in the infrared. © ESO, thanks to Misha Schirmer.

In our Galaxy, it is then observed that the Cross-Scutum-Centaurus arm clearly shows up in the stellar component, while this is not the case of Sagittarius-Carina arm: stars of all ages (except the younger ones) therefore draw a structure with two main arms, the two secondary arms being mainly dominated by gas and very young stars of Population I. The two main arms, Cross-Scutum-Centaurus and Perseus, appear connected to the ends of the central bar that is certainly at the origin of the symmetrical density waves that are the arms.

Structure and components of the Milky Way

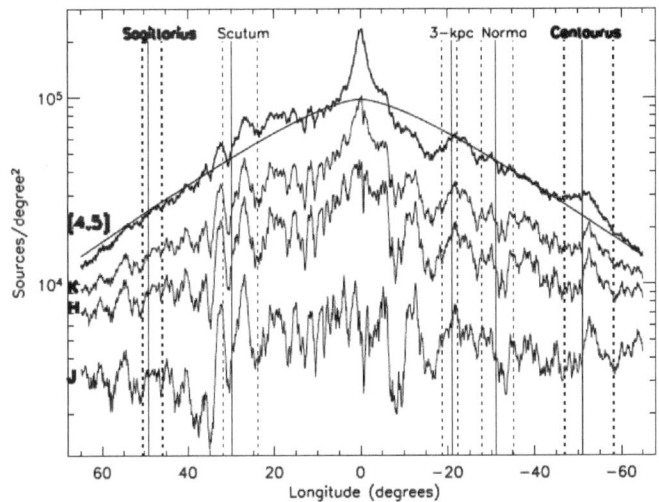

FIG. 3.29 – Profile of the infrared emission of the inner regions of the Galaxy. The number of sources per square degree within a band of 2° wide along the galactic plane is plotted as a function of galactic longitude for different wavelengths: bottom to top, J band (1.2 μm), H (1.6 μm) and K (2.2 μm) (GLIMPSE catalog data) and 4.5 μm (SPITZER satellite data). The longitudes where the spiral arms are seen tangentially are indicated. Here there is no trace of the Sagittarius-Carina and Norma arms, while both sides of the Cross-Scutum-Centaurus arm clearly appear (see Fig. 3.26 for the geometry of the arms). The structure called "3-kpc" for historical reasons is actually a ring around the center of the Galaxy at the ends of the bar. At 4.5 μm, there is virtually no extinction by dust, while the profiles at shorter wavelengths are affected by extinction, which is very irregular. From Benjamin, R.A. (2008) *Astronomical Society of the Pacific Conference Series* 387, 375-380, with thanks.

## 3.6 Dark matter in the Galaxy

As we saw at the beginning of this chapter, the kinematic study of matter orbiting around the center of our Galaxy, in particular the gas, provides information on the total mass in the Milky Way. Although the distances are not very well known beyond the Sun, observations reveal a nearly flat rotation curve up to the last measuring point, as in all other spiral galaxies. While the inclusion of only visible matter (stars and gas in the bulge and the disk) predicts a velocity that would decrease from the Sun (Fig. 3.30), the observation of a flat rotation curve $V_{\rm rot}(R)$ implies the existence of invisible mass, at least beyond the Sun. The mass $M(R)$ comprised within a given radius $R$ then increases linearly with $R$, to a first approximation; in the simplifying assumption that the mass would have a spherical distribution, one would have $GM(R) = RV_{\rm rot}^2$.

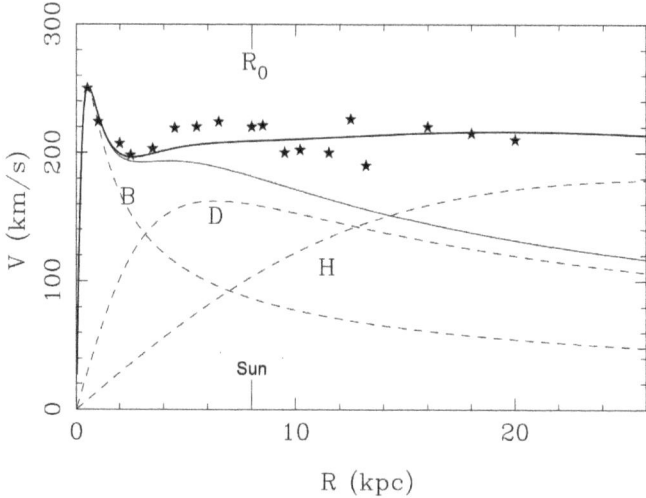

FIG. 3.30 – The rotation curve of the Milky Way, i.e. the rotation velocity as a function of the distance $R$ to the center. The data resulting from observations of the gas are represented by starred symbols (a more elaborate version is presented in Figs. 3.2 and 3.31). The model fit to the data (thick solid line) is composed of a bulge (B) a disc (D) and a spherical halo of dark matter (H), whose contributions are dotted curves. The thin solid line curve represents the sum of the contributions of the visible matter (bulge + disk). The need for a dark matter halo is obvious beyond the solar radius.

In our Galaxy, whose type is SBb in the Hubble sequence, as in all the massive spiral types Sa, Sb and Sc of this sequence, the need for dark matter appears only in the outer parts of the disk. The center of these galaxies is dominated by visible matter. It is however possible that there is dark matter at the center, but its amount is difficult to estimate, due to the uncertainty in the mass-luminosity ratio $(M/L)$ of stars, since the mass of the stellar component is derived from its brightness. This ratio depends of course of the nature of the stellar populations, their age and metallicity. The dispersion in $M/L$ is larger when using the visible light rather than the near-infrared one, but even in the latter case the dispersion can still be up to a factor of 2 to 3. To obtain at least a crude idea of the distribution of dark matter in the center of galaxies, one might consider dwarf galaxies of low surface brightness, which are so dominated by dark matter and the contribution of the stars to the total mass so small that an error of a factor of 2-3 on $M/L$ has little effect on the amount of dark matter inferred from the rotation curve. In these galaxies, the dark matter in their central region has a volumic density roughly constant with radius, and does not exhibit the central peak that is predicted by some numerical simulations of galaxy formation. We can assume that this distribution of dark matter also applies to the inner

# Structure and components of the Milky Way

regions of spiral galaxies. This contradiction is one of the problems faced by the standard theory of cold dark matter (CDM or Cold Dark Matter), which is generally used to account for observations of galaxies.

## 3.6.1 The contribution of baryons

What could be the nature of this invisible matter? Could it contain ordinary matter: protons and neutrons, i.e. baryons? From cosmological observations, we know very well today the amount of baryons in the Universe. Several additional observations provide a body of evidences, which converge to the same value: baryons make up 17% of the whole mass in the Universe, the rest being dark matter of different nature (here we exclude the dark energy that seems to dominate the dynamics of the Universe). The amount of baryons in the Universe is known thanks to the results of the primordial nucleosynthesis, which occurred in the first fifteen minutes after the Big Bang. From the present abundance of the light elements formed in these first moments (Helium, Deuterium, Lithium, Beryllium, Boron), we deduce that the ratio of the density of baryons to the critical density the Universe is 5%. This critical density, which accounts for the zero-curvature Universe we observe, is about $10^{-29}$ g cm$^3$; all the matter makes up 30% of this (5% for baryons and 25% of another nature), the remaining 70% corresponds to the dark energy. These figures have been obtained in recent years through the detailed observations of the cosmic microwave background radiation and its anisotropy by the WMAP and the Planck satellites, through observations of type Ia supernovae that calibrate distances regardless of the expansion of the Universe, and observations of gravitational lenses, which are used to probe the mass.

From the figures above, it can be conlcuded that we still have a large amount of unseen baryons to discover. Indeed, when we sum the amount of baryons identified in galaxies as stars and gas, this corresponds to only 6% of all baryons in the Universe. In rich galaxy clusters, there are large amounts of very hot gas (tens of millions of degrees) that emit X-rays, and there are often more baryons in this gas than in the galaxies in the cluster; but on average in the Universe, this contribution amounts to only 3% of the baryons. A large amount of baryons are also found in the intergalactic medium, in the form of cosmic filaments observed by the absorption they produce in the Lyman α line of hydrogen in the spectrum of distant sources such as quasars: the corresponding fraction may reach 18%. Moreover, 5 to 8% of this gas could be very hot and would not be detectable through Lyman α absorption. The total could reach 35%, but 65% of baryons still remain to be identified.

This census shows that it is quite possible that some of the invisible matter in our Galaxy is made up of baryons. Some time ago, it was considered that the missing baryons might be in the form of white dwarfs, which are difficult to detect. But this hypothesis was quickly eliminated because white dwarfs are stars at the end of their life, which would have previously synthesized and ejected heavy elements in amounts well beyond what is observed. The number in white dwarfs near the Sun is consistent with the predictions of stellar evolution, the corresponding mass fraction is very small and their contribution to the baryonic dark matter is almost negligible.

Another hypothesis proposed that missing baryons make brown dwarfs. Brown dwarfs are "failed stars" whose mass is below the threshold for nuclear reactions. These are very cold and extremely compact objects that emit little radiation. In the 1990s, several groups were engaged in searches for brown dwarfs in the Galaxy, and more generally for various kinds of exotic objects that could contain mass, like primordial black holes. These were the MACHOS (Massive Compact Halo Objects) project, the EROS (Experience pour la Recherche d'Objets Sombres, i.e. Experiment for the search of dark objects) project, and the OGLE (Optical Gravitational Lensing Experiment) project. The method used in these projects was to detect the increase in flux of a star when a mass passes by chance in front of it and acts as a gravitational lens (a well known effect of general relativity, which is referred to as the *gravitational microlensing*, as opposed to the large gravitational lenses that are the galaxies and clusters of galaxies). The microlensing amplification can last a month or two; it is longer if the mass of the dark object is greater. However several unknown parameters are involved in this phenomenon, such as the transverse velocity and the relative distances of the star and the lens, so that we can only obtain statistical results. MACHOS and EROS have focused on stars in the outer Galaxy and in the Magellanic Clouds, and OGLE observed the stars of the bulge of the Galaxy.

The three projects have effectively detected gravitational microlenses, but the number of events observed after more than a decade of observations of millions of stars is too low for dark compact objects to be responsible for the invisible baryonic matter (Fig. 3.31). All observed events are compatible with lenses that are low luminosity stars of the Galaxy or the Magellanic Clouds, or black holes, remnant from the explosion of very massive supernovae: it is not necessary to resort to exotic objects that are not included in conventional stellar populations. In summary, no dark matter object has been detected, as the number of events was that expected with the visible mass alone and the usual values of $M/L$.

It is noteworthy that only very few microlensing events have been identified due to objects belonging to the thick disc of the Milky Way. Yet, as we have seen in section 3.2, the thick disk has a stellar population that is, on average, older than the thin disc, and a larger $M/L$. This confirms that the mass of the thick disk is only a small fraction of the mass of the thin disk and bulge of the Milky Way, although several studies have suggested a higher mass.

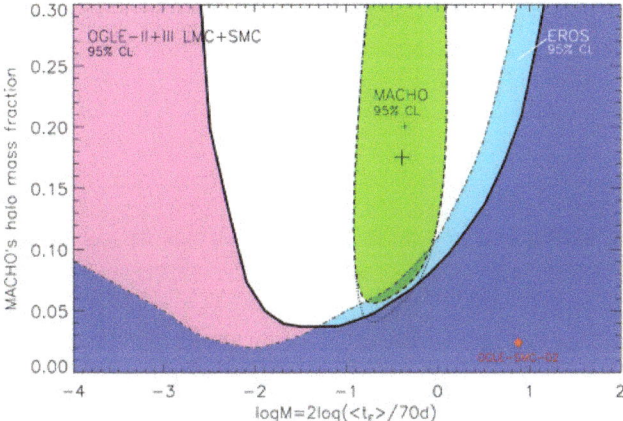

FIG. 3.31 – Upper limits of the mass fraction of the dark matter halo of the Galaxy included in compact objects (MACHOS), as a function of their mass ($x$-axis). The limits are obtained by the observations of the OGLE (pink), EROS (blue) and MACHO (green) microlensing projects. Here, the dark matter is supposed to be distributed in a spherical halo[5]. The colored parts are allowed areas. The red square represents the limit of the mass that would be included in possible black holes, such as might be the object OGLE-SMC-02: there would be 2% of the mass of the halo in this form, which is compatible with the expected abundance of stellar black holes, residues of very high mass stars at the end of life. From Wyrzykowski, L. et al. (2011) *Monthly Notices of the Royal astronomical Society*, 416, 2949-2961, with permission of the publisher.

### 3.6.2 A gas contribution?

If the baryonic dark matter is not made of compact objects, it could exist as a more diffuse form of gas clouds that would not produce gravitational lensing. Cold molecular gas at low temperature ($T = 10\text{-}20$ K) is an excellent candidate to form the baryonic dark matter. We saw in section 3.3.2. that molecular hydrogen is undetectable under these conditions. In the internal disk of the Galaxy, it is possible to trace the molecular gas through the CO molecule. But the abundance of CO relative to $H_2$ decreases with increasing distance from the center, along with the abundance of all heavy elements, as will be discussed in Chapter 7. In the outer parts of the Galaxy, the dense hydrogen clouds where atomic hydrogen turns into molecular hydrogen become invisible in CO lines; while they are totally undetectable directly.

---

[5] This halo is believed to obey a density law $\rho_H(R) = \rho_0(R_0^2 + a^2)/(R^2 + a^2)$ with $\rho_0 = 0.0079$ M$_\odot$ pc$^{-3}$, $R$ being the galactocentric radius, with $R_0 = 8.5$ kpc and $a = 5$ kpc. One would obtain different limits with other models of the distribution of dark matter.

Among the various components of the Galaxy, only the observable interstellar gas, which is dominated by atomic hydrogen, does not decrease exponentially with increasing distance $R$ to the center: its surface density, rather, decreases as $1/R$, like the dark matter (Fig. 3.11). As soon as observations of the gaseous galaxy disks were made, it was noted that the HI gas surface density and that of dark matter, as deduced from the rotation curves, are proportional in the outer parts of galaxies: the mass surface density of the dark matter is between 7 and 10 times that of the HI gas. This applies both to massive spiral galaxies, which have relatively little dark matter, and to dwarf galaxies, where dark mass dominates the dynamical mass. For now, this relationship has no physical explanation accepted by everybody; but it could be understood if the fraction of gas in dense molecular clouds was constant everywhere.

### 3.6.3 Distribution of dark matter in the Galaxy

It is often assumed, for simplicity, that the dark matter halo is spherically distributed around the Galaxy. Yet the standard cosmological model simulations predict that dark matter should instead form a flattened ellipsoid. How can we obtain information about this distribution? Is there dark matter in the disk of the Galaxy?

We have seen that the gas disk, as traced by the 21-cm line of atomic hydrogen, flares strongly in the outer parts (Fig. 3.12). The variation with radius of the thickness of the gas is of considerable interest for the knowledge of the distribution of mass in the disk, at least in principle. Indeed, it may be assumed that there is hydrostatic equilibrium between the gravitational attraction towards the galactic plane and the "pressure" corresponding to the macroscopic motions of the gas, whose velocity dispersion perpendicular to the galactic plane is 6 to 7 km/s. Observation of the 21-cm line in external galaxies seen face-on shows that this dispersion does not vary from galaxy to galaxy, and is not a function of radius. This is not surprising, because the gas is dissipative and has no difficulty in radiating its surplus energy: a dynamic equilibrium occurs between heating, which is dominated by gravitational instabilities in the outer regions of the disk, and radiation. We can therefore deduce the radial variation of the gravitational attraction of the disk from the thickness of the gas disk. The problem, however, is complicated by the existence of a poorly known additional pressure due to cosmic rays and to the magnetic field, which is probably low in the outer disk, but makes the results somewhat uncertain.

Studies of the vertical equilibrium of the gaseous disk can therefore give us an idea of the 3-dimensional distribution of dark matter, especially the flattening of the dark halo or the fraction of dark matter included in the disc. The result is that the disk of the Milky Way near the Sun is self-gravitating with little dark matter, and that the dark matter within the disk has a total parameters $M = 2\text{-}3 \times 10^{11}$ M$_\odot$, and includes a ring between $R = 13$ and 18.5 kpc. The existence of this ring is confirmed by the observations of gamma rays by the satellite GRO, which show a similar concentration of sources of gamma rays in the outer parts of the Galaxy (Fig. 3.32). The gamma-ray emission could either come from the interaction of cosmic rays with the molecular material, or, as suggested by the authors of the study, from the decay of any exotic particles that would constitute the dark matter (the Weakly Interacting Massive Particles, or WIMPs). The ring also coincides with a giant stellar structure that surrounds the galaxy: the Monoceros Ring, discovered in 2002 using the Sloan Digital Sky Survey (SDSS), which contains some $10^8$ M$_\odot$ of stars. It could be that this stellar structure corresponds to the remains of the interaction with a satellite, which would have formed some of the tidal arms wrapping around the Milky Way. The stars would be debris of this companion, and a part of its dark matter would also be wrapped around the galaxy to form the ring.

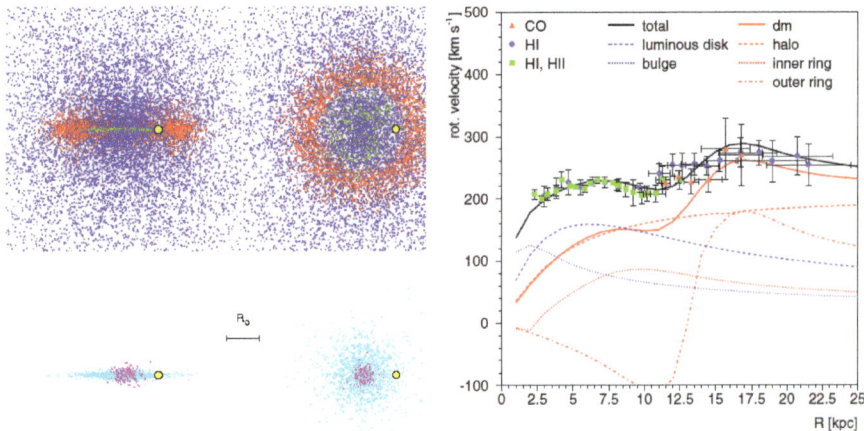

FIG. 3.32 – Left: a model of the distribution of dark matter in the Galaxy (top, blue for the halo, red for the ring) and of visible matter at the same scale (bottom, pink for bulge, green for the disc). The Milky Way is seen edge-on and face-on. The location of the Sun is the yellow circle and the distance from the Sun to the galactic center (taken here of 8.3 kpc) is indicated by a bar. The distribution of dark matter is partly derived from gamma ray observations with the satellite GRO (EGRET experiment). Right: the rotation curve, where the symbols with error bars are the measuring points corresponding to the observations of CO, HI, HII as shown on the top left of the diagram. This curve is identical to that of Figure 3.2, the rotation velocity at the Sun being taken as 220 km/s. The curves show the rotation curves that would be predicted from each of the different components of the mass of the Galaxy if it was alone (dm for dark matter, and corresponding contributions in red). The black curve results from their combination. Note that the contribution of the outer ring to the rotation is negative at 10 kpc, as it exerts an outward force on the interior regions. From de Boer, W. et al. (2005), *Astronomy & Astrophysics* 444, 51-67, with permission of ESO.

One of the main problems of the standard theory of dark matter is the lack of detection of potential candidates for the corresponding particles. These candidates are not included in the standard model of elementary particles, and to explain their possible existence we must make assumptions yet unverified, such as those of the super-symmetry theory. In this theory, it is assumed that each existing particle has a super-symmetric counterpart. The most stable of these new particles would be the neutralino, one of the best candidates for the WIMPs, which are supposed to interact extremely weakly with other matter, which explains why they have not been detected until now.

### 3.6.4 An alternative possibility: modified gravity

Another solution to the problem of the flat rotation curves was proposed in 1983 by Mordehai Milgrom: the idea is to abandon dark matter, and instead to modify the law of gravity in the weak field regime, such as the outer parts of galaxies. This hypothesis was called MOND (for MOdified Newtonian Dynamics). Indeed, the apparent need for dark matter is present in different astrophysical contexts when the acceleration of gravity falls below a critical value $a_0 = 2 \times 10^{-10}$ m s$^{-2}$ (of the order of one angström per square second). It is then assumed that below this value, the acceleration $g$ due to a point source of mass $M$ located at the distance $R$ starts to depart from the Newtonian value $g_{\mathrm{Newton}} = GM/R^2$, to tend to the square root of the product of $g_{\mathrm{Newton}}$ by $a_0$: $g = (a_0 \, g_{\mathrm{Newton}})^{1/2}$.

In these conditions, the gravitational force is no longer proportional to the mass $M$, but varies as the square root of the mass; also, it now decreases as $1/R$ and not as $1/R^2$. This last point automatically explains why the rotation curves are flat very far from the center of galaxies: the value of the rotation velocity is then such that $V_{\mathrm{rot}}^4 = GM a_0$, where $M$ is the total mass of the visible galaxy.

The success of this hypothesis for nearby galaxies is remarkable. Virtually all rotation curves can be explained without dark matter and with no free parameter: the only parameter is $a_0$, which is a universal constant. Even dwarf galaxies, which seem dominated by dark matter into their central parts, share this relationship. For them, the acceleration of gravity is already less than $a_0$ near the center. These galaxies are laboratories to test this new law of gravity, which we cannot check on the Earth: the gravity to which we have access is fully Newtonian, as is the gravity in the Solar system. We would have to go to a distance of about 8000 times the Earth-Sun distance from the Sun to hope to reach $a_0$; but at this distance, the gravity of the Galaxy itself, which is greater than $a_0$, would have to be taken into account.

The variation with radius of the rotation curve of the Milky Way could help us to better understand the transition between the Newtonian regime and the "MONDian" regime. Presently, this transition is not well known;

several transition functions are predicted by various theories that are not yet well established. The more precise determination of the distance and the kinematics of stars in the Galaxy with the future astrometry satellite GAIA should allow some progress in this domain.

One of the successes of MOND is to predict the Tully-Fisher relation between the rotation velocity of spiral galaxies and their brightness, discovered in 1978 by the American astronomers R. Brent Tully and J. Richard Fisher. This relationship was improved over time, and the brightness was replaced by the visible mass, which includes not only the stars but also the gas, especially for dwarf galaxies that are dominated by gas and (apparently) by dark matter. Figure 3.33 presents a recent version. The observed relationship is such that the visible mass of galaxies varies as their rotational speed to the fourth power, which is exactly the prediction of MOND. Conversely, the standard model does not predict the correct slope of the relationship nor its zero point. To explain why so few baryons condense in galaxies, one has to assume that the gas is largely expelled by supernovae, especially in dwarf galaxies that have a low potential well, so that the escape velocity is reached easily. For massive galaxies, one also invokes the effect of their central black hole, whose energy could prevent the formation of stars and condensation of baryons.

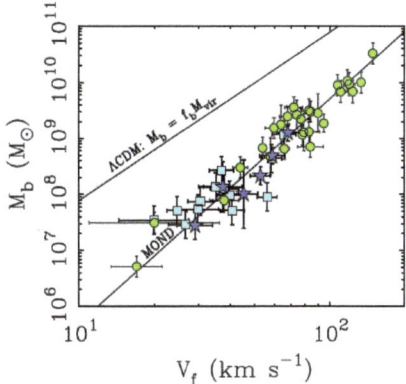

FIG. 3.33 – The Tully-Fisher relation between the visible (baryonic) mass of a galaxy and the rotation velocity in the outer parts of the galaxy, where the rotation curve is flat. Of course, this velocity has been corrected for the inclination of galaxies on the line of sight. Massive galaxies like the Milky Way are at the top right (V > 150 km/s). The blue squares correspond to gas-rich dwarf galaxies, in which most of the visible matter is gas, but which are apparently dominated by dark matter. The straight line fitting the observational points has the slope 4. The prediction of the Cold Dark Matter standard model is the line above, with the slope 3. From McGaugh, S.S. (2011) *Physical Review Letters* 106, 121303 (http://cdsads.u-strasbg.fr/abs/2011PhRvL.106l1303M) with permission of the American Physical Society.

# Chapter 4
# The galactic center

The center of our Galaxy hosts a rather modest bulge, which has the shape of a dumbbell, or rather more similar to a peanut. This kind of bulge often forms directly from the stellar disc when there is a strong bar in a galaxy. It contains a black hole in its center. We will see that the mass of the super-massive black holes at the center of galaxies is proportional to the mass of their bulges: the black hole of the Milky Way has thus a low mass. However, it is of major interest because it is the closest super-massive black hole that we can observe, and for which we have very accurate data.

## 4.1 Bar and bulge

Our galaxy is a barred spiral, like the majority of spirals in the Universe. Yet the presence of the bar was established only recently, thanks to near-infrared imaging, which avoids the problem of dust extinction. In the 1970s, we already had a kinematic indication of the existence of the bar, from the observation of highly non-circular gas motions. Indeed, the observations of gas in the HI and CO lines reveal, through the Doppler effect, the motions of the gas in the center of the Milky Way: they have strong positive and negative velocities, over 200 km/s, while zero radial velocity is expected. Some astronomers have tried to explain these velocities by a rapid gas ejection in the vicinity of the black hole, but the later is not active and cannot be responsible. Put more simply, gas orbits are ellipses whose major axis is aligned with the bar: thus the gas comes towards us from one side of the galactic center and moves away from the other; the projection of the velocity on the line of sight is close to the rotational velocity. In the 1990s, mapping the whole sky in the near infrared, the COBE satellite showed directly the existence of the bar. Viewed from a distance of 8 kpc from the center, the position of the Sun, the bar appears to us in perspective. It has about 4 kpc radius, and is inclined by 20° to the line of sight: the side of the bar located at positive longitudes is significantly closer to us than the opposite end of the bar, at negative longitudes (Fig. 4.1).

In the 2000s, the exploitation of star counts in the near infrared, from the whole sky imaging survey 2MASS (Two Micron All Sky Survey), confirmed the existence of the bar, showing an asymmetry in the star number between positive and negative longitudes, on either side of the center. On the other hand, these counts suggest the existence of a nuclear bar, which would be embedded in the main bar; it would have a radius of 150 pc, and be almost perpendicular to the main bar. The existence of such a secondary bar is very common in close spiral galaxies. The numerical simulations have shown that such nuclear bars develop spontaneously. They occur in a fairly advanced stage of bar development in a spiral galaxy. In an axisymmetric disk, where the velocity dispersion of stars and gas is low enough to make the disk unstable, "primary" bars first begin to develop. Their size is about half the radius of the visible galaxy. The bar rotates at an angular pattern speed constant with radius, which places the co-rotation resonance just at the end of the bar, thus the middle of the stellar disk. Inside co-rotation, stars rotate faster than the bar.

During its evolution, the mass is gradually concentrated towards the middle of the bar, which speeds up its rotation. In fact, the bar has a non-axisymmetric morphology, which generates tangential gravity forces and therefore torques. These torques produce an exchange of angular momentum between stars and gas. The later is driven towards the center, and thus the mass distribution is increasingly concentrated. There comes a critical moment when the speed of the main bar is no longer sufficient to stabilize it against gravity, and a secondary bar decouples, which rotates faster around the center.

As the two bars do not rotate at the same speed, their relative orientation is arbitrary: they may be sometimes parallel and sometimes perpendicular. They are not completely independent, however, as they exchange energy at resonances. Most of the time, in the simulations, the co-rotation of the secondary bar corresponds to the inner Lindblad resonance of the primary bar (we will see in Chapter 5 what this resonance is). As a gas ring in general forms at the inner Lindblad resonance, one expects to find the nuclear bar exactly included inside the ring. This is probably what also happens to the Milky Way: there is indeed a molecular gas ring about 200 pc in radius.

This secondary bar has an important role in galaxies: it prolongates the action of the primary bar to the vicinity of the nucleus. In the first phase of the evolution of the primary bar, when it exists alone, the galaxy gas falls toward the center, but accumulates at the inner Lindblad resonance (ILR). For the gas to continue to fall toward the center, we must break the axial symmetry, which requires another strong bar inside. Our Galaxy is going through this phase, and the secondary bar will soon transport the molecular ring gas toward the central black hole. Our nucleus is still inactive today, but will become active when supplied with gas.

The galactic center

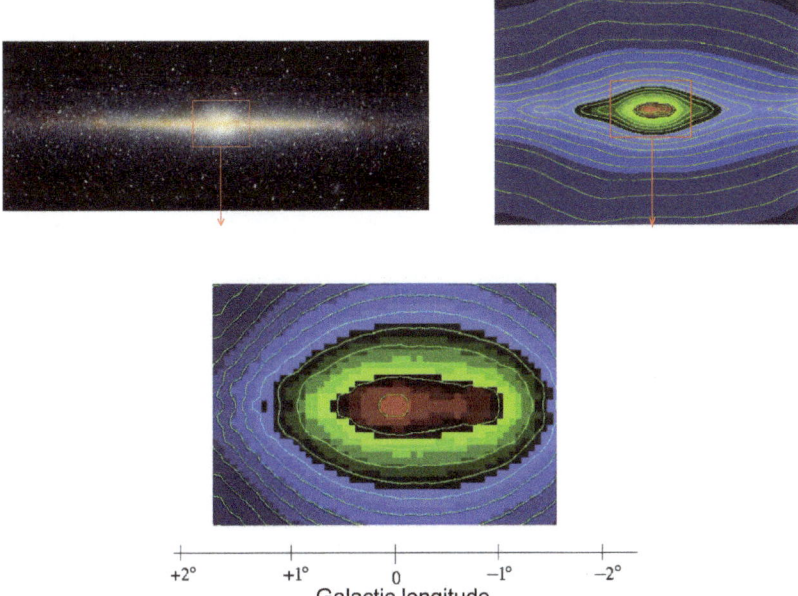

FIG. 4.1 – Top: a view of the Galaxy obtained by the COBE satellite in the near infrared. The longitude goes from −180° at the right to 180° at the left. Middle, zoom of the central part, from −10° to 10° in longitude. Bottom: zoom again, from −1.5° to 1.5° in longitude. From Alard, C. (2001) *Astronomy & Astrophysics* 379, L44-47, with permission of ESO.

## 4.2 The interstellar matter at the galactic center

While the molecular gas is very abundant in a ring between 4 and 8 kpc from the center, it seems deficient inside. Yet when we look to the central regions within a radius less than 0.5 kpc, the emission of the CO molecule, the molecular hydrogen tracer, reveals a pronounced maximum. This corresponds to the molecular ring of 200 pc radius we have mentioned, a very dense ring in which the nuclear bar is inscribed. This region is called the "Central Molecular Zone" or CMZ (Fig. 4.2). The ring emits many molecular lines, some of which are high-density tracers as CS or HCN. Some lines correspond to transitions between high rotational levels, indicating that the gas temperature is 50 to 70 K.

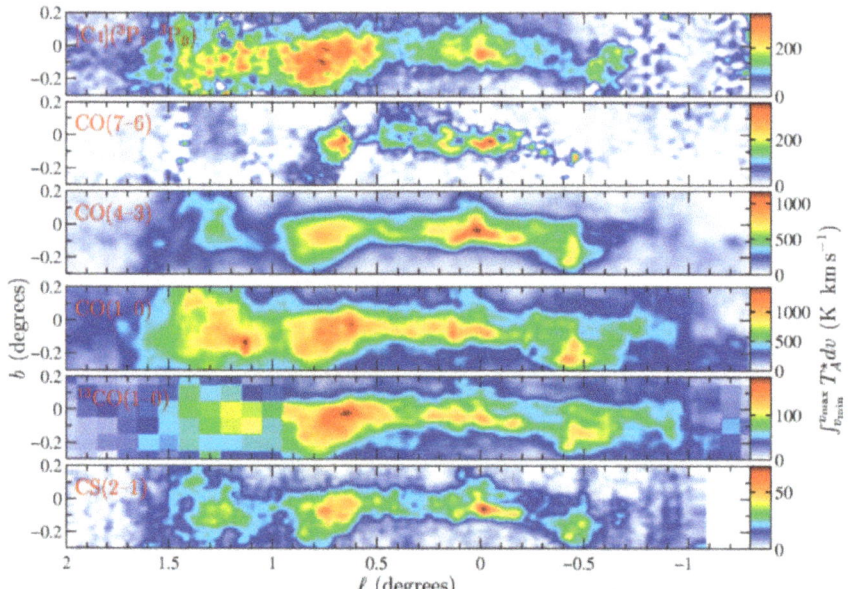

FIG. 4.2 – Maps of the galactic center, including the central 3° in longitude and 0.5° in latitude, showing the molecular ring of the CMZ (Central Molecular Zone). The three top maps were obtained with the sub-millimetric telescope AST/RO of 1.7 m at the South Pole. The neutral carbon line CI at 492 GHz traces the gas at the surface, while the CO J=7-6 line at 807 GHz is dominated by the warm and dense gas. The three bottom maps at lower frequency were obtained with the Bell Laboratories telescope of 7 m. The CO J=1-0 line at 115 GHz traces the gas at the surface, while that of $^{13}$CO J=1-0 at 110 GHz the gas at larger depth, and the CS J=2-1 line at 98 GHz traces the dense gas. From Martin, C.L. et al. (2004) *Astrophysical Journal Supplement* 150, 239-262, with permission of the American Astronomical Society.

The molecular ring is animated by important non-circular motions, probably due to the bar. Moreover it is not symmetric: 75% of the emission is from the positive longitudes, on only one side as seen in Figure 4.2. The position-velocity diagram of this region shows a characteristic parallelogram structure (Fig. 4.3). In such a position-velocity diagram, a "normal" component animated by only circular motions, and distributed symmetrically with respect to the center, would have a narrow emission passing through the origin ($l = 0$, $V = 0$).

The parallelogram reveals velocities that are prohibited in pure rotation. They can be explained by the presence of elongated orbits along the bar. As discussed in Chapter 5, families of specific orbits exist in a barred gravitational potential; the main families are orbits parallel to the bar (×1 orbits) and perpendicular to it (×2 orbits). The later are present only between the two inner Lindblad resonances, if they exist. The parallelogram feature corresponds to the orbits elongated in the direction of the bar. The fact that the interior of the parallelogram is filled implies the existence of perpendicular orbits, and therefore those which lie between the Lindblad resonances. The center of the diagram is complex, which could be explained by the presence of orbits parallel to the embedded nuclear bar.

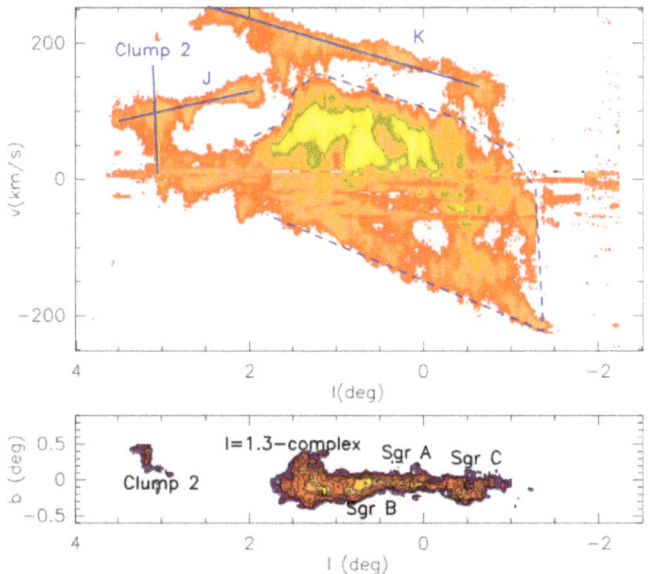

FIG. 4.3 – Longitude-velocity diagram of the galactic center region in the CO(1-0) line, tracing the molecular gas (top), and the same emission in the longitude-latitude diagram. Some of the characteristic molecular clouds are indicated, such as Sgr A and Sgr B. The dashed lines comprise the bulk of the emission, and represent a parallelogram. From Rodriguez-Fernandez, N.J. & Combes, F. (2008) *Astronomy & Astrophysics* 489, 115-133, with permission of ESO.

Unexplained phenomena remain, such as the asymmetry and the offset, which shift the majority of the gas to positive longitudes. It is possible that this shift is a spontaneous gravitational instability. Yet these instabilities are usually damped, and we must assume a recent trigger, either the in-fall of a giant molecular cloud from a larger distance from the center, or the interaction with a companion, a dwarf galaxy orbiting the Milky Way.

Another anomaly is observed in the atomic gas near the center, which supports the hypothesis of a recent disturbance: the central disk seems inclined to the main plane of the galaxy by an angle of 25 degrees. Figure 4.4 shows schematically the observed distribution. The nuclear disc would then have a precession motion about an axis perpendicular to the disk of the galaxy, while it deforms as a function of time. The rotation period at this radius is about 30 million years, and the precession period could be of the same order.

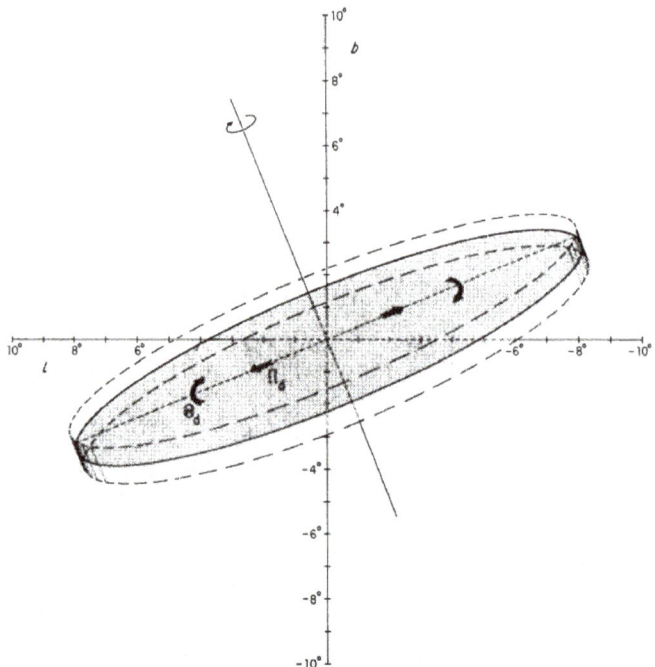

FIG. 4.4 – Schematic representation of the atomic gas distribution in the galactic center regions, as a disk inclined by 25° with respect to the galactic plane. This inclined disk is 1.5 kpc in radius. From Burton, W.B. & Liszt, H.S. (1978) *Astrophysical Journal* 225, 815-842, with permission of the American Astronomical Society.

## 4.3 The black hole

### 4.3.1 The close environment of the black hole

The central parsec of the Galaxy contains two dense stellar clusters: a cluster of old and relatively cold stars of total mass $10^6$ M$_\odot$ and a cluster of very young and warm stars in the center of total mass $1.5 \times 10^4$ M$_\odot$. Among these young stars, two streams differ in their kinematics, one rotating in the same sense as the Galaxy, the other in counter-rotation. These two systems are arranged in two thick disks that make a wide angle between them. The cluster of old stars has a quasi-isotropic velocity distribution, and rotates more slowly as a solid body.

The gas in the very central parts of the Galaxy has a very complex structure. Its study is of great interest to better understand the physics in the vicinity of supermassive black holes, because the black hole of our Galaxy is the one for which we have the most information, by far. This black hole is first surrounded by a central cavity (CC), of dimensions 2 × 3 pc. This cavity was probably emptied by stellar winds but still contains some gas photo-ionized by the ultraviolet radiation from the young star cluster mentioned earlier. In this cavity one can clearly distinguish a mini-spiral, composed of three spiral arms, with a small central bar (Fig. 4.5). This mini-spiral corresponds to photo-ionized gas falling onto the black hole, which is torn and stretched by tidal forces, and forms spiral arms by differential rotation. The total mass of ionized gas in this mini-spiral is estimated at 60 $M_\odot$, while there would be about 200 $M_\odot$ of diffuse ionized gas distributed throughout the cavity. A ring of dense molecular gas surrounds the cavity, the "Circum Nuclear Ring", or CNR. It is quite asymmetric, and certainly in full dynamical evolution. Its average particle density is $7 \times 10^4$ cm$^{-3}$, and its temperature 300 K. Its total mass is estimated at $10^6$ $M_\odot$, and it is the main gas reservoir of the central black hole. It can be considered as an accretion disk that would be settling into equilibrium with deformations and oscillations of its plane.

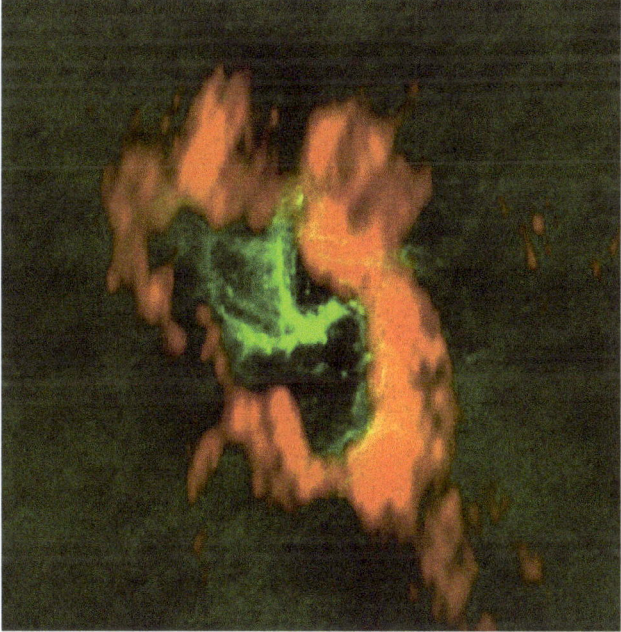

FIG. 4.5 – Image of the galactic center showing in green the radio continuum emission at 3.6 cm wavelength, tracing the ionized gas distributed in a 3-arm mini-spiral, and in red the HCN molecular emission in its J=1-0 rotational line, showing the circumnuclear ring (or CNR). This circumnuclear ring has an outer diameter of 6 pc. It is inclined by 70° with respect to the plane of the sky, and by 20° with respect to the galactic plane. From Farhad Yusef-Zadeh, reproduced in Ferrière, K. (2012) *Astronomy & Astrophysics* 540, 50, with permission of ESO.

Around these two structures, we observe particular regions of ionized and molecular gas, which are probably related to a recent episode of star formation. First, an ionized gas shell exists, 10 pc in diameter, which emits a synchrotron radio continuum, showing that it is full of high-energy electrons. This is a supernova remnant (SNR). It is included in a more diffuse halo of ionized gas 20 pc in diameter. Finally, there is a belt of massive molecular clouds that stretch up to 30 pc along the Galactic plane; some well-known giant molecular clouds are found here, such as the East Cloud (or EC) and South cloud (or SC) that can be seen in the diagram of Figure 4.6.

In the cavity and along the mini-spiral, dust filaments were detected in the infrared. By observing them several years apart, one could detect their proper motions, showing that their kinematics does not match a simple rotational motion around the black hole; Conversely, some appear to be ejected, as if a flow came from the galactic center, either from the cluster of young stars and their stellar winds or as jets from the vicinity of the black hole, or both. Confirming these divergent streams, small very compact clouds, with comet-like morphologies, were also observed. A reconstruction of the 3-dimensional structure has been made for the mini-spiral and the cavity, showing that the three spiral arms are actually streams that are not in the same plane. Some are in simple rotation and could feed the black hole. In addition to the ionized gas, there is also 300 $M_\odot$ of atomic gas in the cavity.

The massive molecular clouds that form the cloud belt have masses of about $10^6$ $M_\odot$, and probably are the links connecting the CNR ring to the molecular medium at a larger scale (100 pc) described earlier. It is likely that the molecular gas settles in several reservoirs of different scales, before gradually losing its angular momentum and finally be able to feed the black hole. Typical emissions in shocks, such as the vibrational lines of molecular hydrogen in the near infrared, were detected in the gas belt, showing the violence of the gas in-fall and collisions in these regions.

The picture that emerges from these complex structures in the galactic center is one of competition between two opposing processes: gas accretion by the black hole and gas ejection by stellar winds, caused by the young star cluster in the nucleus, and also a recent supernova explosion which formed a shell of ionized gas (SNR).

### 4.3.2 Flares near the black hole

The Galactic center source, referred to as Sgr A*, emits intermittent bursts of radiation in millimeter radio waves, in the infrared or in X-rays. When detected, these bursts repeat themselves with a quasi-period of about 20 to 40 minutes. This very short period already gives an idea of the distance from their source to the black hole, whose mass is about $4 \times 10^6$ $M_\odot$.

*The galactic center*

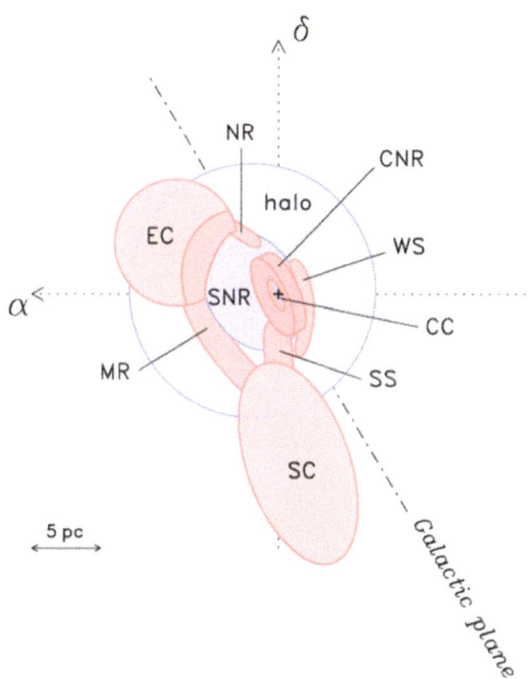

FIG. 4.6 – Schematic representation of the morphology and relative disposition of the different gas structures in the Galactic center, and more precisely within the central 10 pc. This view is the sky projection, as observed from the Sun. The Galactic plane is indicated by the diagonal dot-dashed line ($\alpha$ and $\delta$ are the coordinates of right ascension and declination in the sky plane). In blue are the diffuse structures, essentially ionized gas: the central cavity (CC), the supernova remnant (SNR) and the halo. In pink, the denser molecular gas, including the circum-nuclear ring (CNR) and the series of giant molecular clouds along the molecular belt (EC, SC), and also the Molecular Ridge (MR), the North Ridge (NR), etc. The cross is at the black hole position. From Ferrière, K. (2012) *Astronomy & Astrophysics* 540, 50, with permission of ESO.

The dynamical time (revolution period) is of this order of magnitude at a distance of 0.2 astronomical units (1 AU is the Earth-Sun distance) of the black hole. This distance corresponds to about twice the Schwarzschild radius $R_s$, or horizon of the black hole, the limiting radius below which the relativistic effects dominate, and the point of no return for matter and light. The physical mechanism at the origin of these flares is still unclear. Several models have been proposed: adiabatic expansion of a plasma concentration emitting synchrotron radiation, electron heating in a jet, or emission by a rotating clump of matter in the vicinity of the last stable orbit around the black hole (i.e. at a few $R_s$), the flare period corresponding to one revolution. This clump of matter would be heated by magnetic effects: the highly magnetized accretion disk of the black hole being in differential rotation, the magnetic field reconnects, releasing energy inby shocks.

### 4.3.3 The black hole itself

In the center of the Galaxy, we find the powerful radio source Sgr A*. Although this is not an active nucleus such as is observed in some galaxies, the source Sgr A* is bright enough to be studied by very long baseline interferometry (VLBI), with a very high spatial resolution. The size of the source has been found to be 3 to 10 light-minutes. Such a power in such a small volume can only be the result of the presence of a black hole, thanks to its efficiency to transform mass energy into radiation when matter is falling into it. The demonstration of the existence of a super-massive black hole in Sgr A* and the measure of its mass has come from observation across several years of about thirty stars orbiting directly around it. Their velocities are on the order of 1000 km/s, and some are approaching the black hole at a distance of about a dozen light-hours. Their orbits are Keplerian, centered on Sgr A*, which does not move, as confirmed with great precision by the VLBI proper motion studies. These stars have been followed in the near infrared by two teams of astronomers, on the Keck telescope in Hawaii, and the ESO VLT in Chile. These instruments are equipped with adaptive optics systems that correct images from the atmospheric turbulence. The observations and the derived stellar orbits are illustrated in Figure 4.7. The black hole mass thus obtained is equal to 4.3 million solar masses.

The observations of stars around the black hole raise some questions. First, around a super-massive black hole, the stars should move toward a dynamical equilibrium, with a very high concentration near the center. However, since it takes a long time to achieve this equilibrium, it is expected that the oldest star cluster is more concentrated than the young cluster, which is not the case. Besides, if the relaxation is due to collisions between stars, which are beginning to play a role near the center, one expects a mass segregation, with the most massive stars gathering quickly in the center. However, the most massive stars are normally young stars, which have no time to gather towards the center. Now, we observe that the young, hot and compact stars occupy the central light-month, while the old stars are not so concentrated. An explanation for this paradox is that perhaps the red giant stars of the asymptotic branch, which are more massive and younger than most of the stars, have lost their envelope through collisions; since their hot nucleus is bare, we observe them with a higher temperature and therefore a bluer color.

# The galactic center

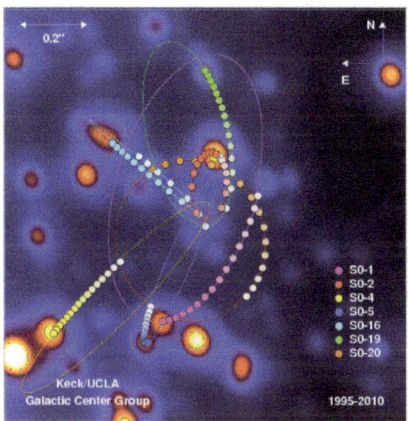 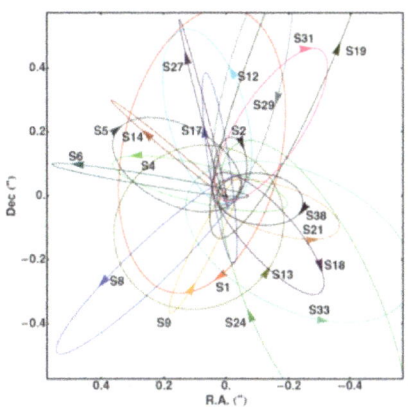

FIG. 4.7 – The stellar orbits around SgrA* have been patiently observed in the near infrared from their proper motions, during the years 1990-2000 by two groups of astronomers, American and German (telescope Keck and VLT-ESO). Left, the map shows how the stellar positions trace from year to year keplerian orbits; the star S2 has a period of 15 years (from Andrea Ghez et al., University of California at Los Angeles, with thanks). Right, the ensemble of stars with their retrieved orbits. From Gillessen, S. et al. (2009) *Astrophysical Journal* 692, 1075-1109, with permission of the American Astronomical Society.

There is still undoubtedly a very young and massive population of stars in the Galactic center. The O-type stars and Wolf-Rayet stars were born 6 million years ago, and only during a 2 million year period. It appears that only massive stars were formed then. One can imagine that a cloud of gas fell to the center 6 million years ago, was trapped in the form of a rotating disk around the black hole, and fragmented by gravitational instability to form massive stars. The problem with this hypothesis is that the gas disk must be very dense to avoid being disrupted by tidal and shear forces. But then, the theory indicates that mostly low-mass stars are expected to form. It might be, however, that the violent gas in-fall, colliding with other clouds that have already started spiraling around the black hole, changed these conditions, which then becomes more favorable to the rapid formation of massive stars, as shown in the simulations of Figure 4.8. The gas is indeed heated to a high temperature by shocks, which favors the formation of massive stars.

Another possibility is that stars are formed in clusters far from the center. Then, by dynamical friction (see the box E4.1), these clusters lose energy and spiral into the nucleus. Arriving at the center, subject to intense tidal forces, they would have been dislocated and redistributed in widely spread stars. However, if the cluster is not sufficiently dense, it is disrupted by tidal forces before arriving at the center, where the massive stars are observed today. This is a real problem because the density that the cluster must have to survive is 100 to 1000 times greater than the density of the stellar clusters that are observed today near the galactic center. To solve this problem, the astronomers thought that the cluster in question had an intermediate mass black hole in its center ($10^4$-$10^5$ $M_\odot$). This black hole could have been formed in the core collapse of a globular cluster, for example.

Finally there are other possibilities to bring massive stars down to distances from the black hole of the order of a light-month. The interaction of binary stars with the black hole could lead to the capture of a star by the black hole, while the stellar companion is ejected to infinity. The gravitational interaction and the scattering produced by the giant molecular clouds could also help to bring massive stars to the center. Other possibilities are collisions between stars. All these processes predict the existence of high velocity stars in the Galaxy, and these are indeed observed in the stellar halo.

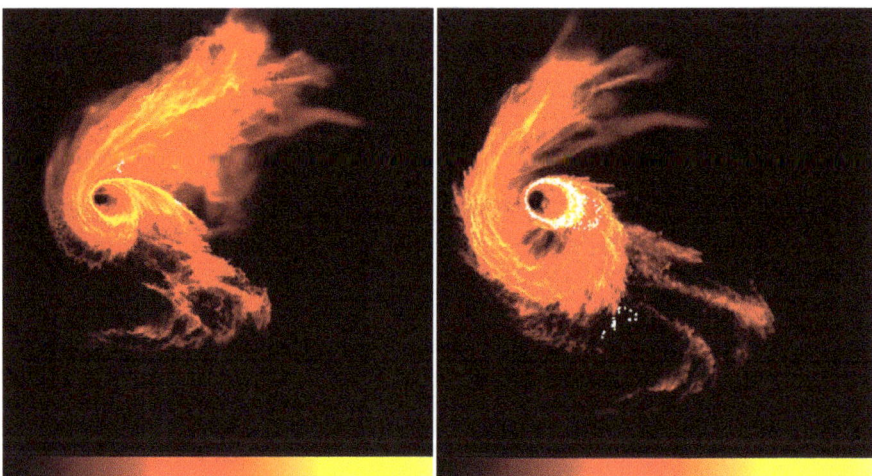

FIG. 4.8 – Simulation of the evolution of a gas cloud of $10^4$ solar masses, falling onto a black hole similar to the one in our Galaxy. The colors indicate the gas density. The size of the box is 0.5 × 0.5 pc, or 1.5 × 1.5 light year. The gas captured around the black hole fragments into stars (white dots), with orbits of large eccentricity, such as observed. Thanks to Ian A. Bonnell.

## E4.1. The dynamical friction

When two stars "collide", i.e. pass close to one another with a high relative velocity, their trajectories are hyperboles (in the center of mass reference frame), and they recess after their encounter with a velocity equal to their initial one: only the direction of their motion has changed in this reference frame. Even if there is no loss of total energy, there is an exchange of energy from one star to the other in the rest frame. If one of the stars for example was at rest initially, the encounter will speed it up and put it into motion. But the effect of these collisions is still very small, because the stars are never very close to each other in a galaxy. One can calculate the two-body relaxation time for a set of stars, which is the time it takes for a given star to lose a significant fraction of its initial energy. This time is greater than the age of the Universe by about 7 orders of magnitude. We can therefore neglect collisions between stars. Note that the two-body relaxation time is even greater in a galaxy, because a large number of stars exist there, and therefore the gravitational potential is softened. A star is not scattered by any neighboring star, because it feels mainly the smoothed gravitational field of the ensemble of N bodies, N being of the order of 200 billion!

The situation is different for a very massive body moving in a sea of stars (Fig. E4.1). Consider, for example, a globular cluster of mass approximately one million solar masses, crossing a galaxy at high velocity. The cluster attracts by its gravity nearby stars, whose density increases locally. Progressively, the diverted stars accumulate in the wake of the globular cluster in motion. The accumulation of mass in its wake slows down the globular cluster. Thus the medium behaves as having a certain viscosity, although there is no contact between the stars, and this produces what experts call a dynamical friction. It is easy to see that the slowed down massive body can lose enough energy to be captured by the galaxy. Chandrasekhar in 1943 has calculated the magnitude of dynamical friction for globular clusters in the Galaxy. But he has not considered that the dynamical friction could act at a distance, and his formula gave a force proportional to the local density of stars.

Since then, it has been possible to calculate precisely, from numerical simulations, the amplitude of the dynamical friction between two interacting galaxies that do not interpenetrate. Tidal forces at large distance deform the galaxies, and these deformations produce excesses of density that can slow down the galaxies in their relative motion around each other. Somehow, the deformations of the galaxies require potential energy that comes from the relative orbital energy of the two galaxies, so that they spiral then towards each other, losing their relative energy, and may eventually merge.

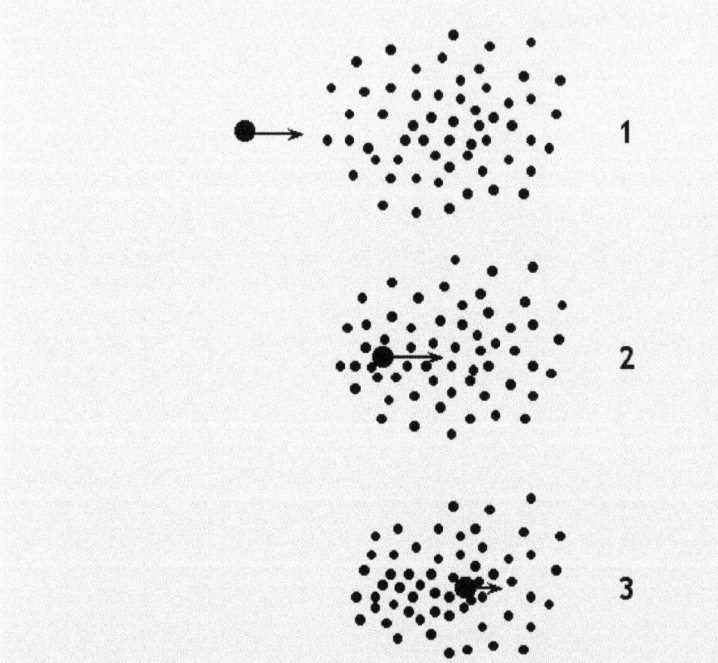

Fig. E4.1 – Principle of dynamical friction. 1: a massive body enters in a sea of stars, whose individual mass is negligible compared to the body mass; 2: it attracts the stars of the galaxy, which tend to accumulate and form a wake behind the body; 3: the mass accumulation in the wake slows down the massive body.

### 4.3.4 Gas in-falling onto the black hole

Up to now, we have considered only the motion of stars around the central black hole. Recently, astronomers have found that one of the faint objects that rotates with a velocity of 1700 km/s around the center was not a star, but a diffuse cloud with a temperature of about 500 K, much lower than the surface temperature of a star. In addition, recombination lines of the hydrogen atom (Brackett lines) were detected in its spectrum, which show that it is indeed ionized gas mixed with dust that radiates in the infrared. This cloud is probably due to stellar winds emitted by young stars that orbit Sgr A*.

The mass of the cloud would be about $10^{-5}$ $M_\odot$ or 3 times the mass of Earth. When discovered, it was heading towards the center of the Galaxy: its orbit, with a period of 137 years, is such that its pericenter happened in 2013-14, when it passed close to the black hole, at a distance equal to

*The galactic center*

3000 times its horizon, or 36 light-hours. Only the stars S2 and S14 in Figure 4.7 came down to distances as close to the black hole. The radial velocity of the cloud has increased from 1200 km/s in 2004 to 2350 km/s in 2011 (Fig. 4.9). Its emission is no longer that of a point source as it was when it was discovered, but extends in the direction of the orbit, since the cloud is stretched under the intense tidal forces from the black hole.

The fate of this cloud around Sgr A* in the coming years is generating a lot of excitement. Figure 4.10 shows a simulation of the evolution of the cloud during and after its closest passage to the black hole. It could happen that X-ray flares are triggered if some part of the gas cloud comes into close proximity of the horizon, on the last stable orbit. While the black hole of the Milky Way remained so far "silent" during human memory, it might wake up as a result of the accretion of gas, leading to powerful radiation if the cloud is completely torn apart by tidal forces and partly swallowed by the black hole. Note however that Sgr A* will not become a real active nucleus, given the low mass of the gas cloud and the fact that most of it will come out unscathed from the pericenter. There will probably be a shock wave, and the cold gas will be heated to about ten million degrees, and emit X-rays.

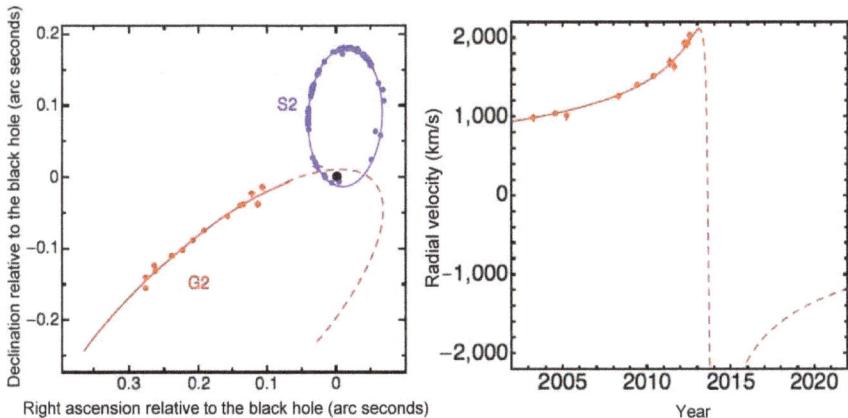

FIG. 4.9 – Left, the gas cloud orbit around the black hole (in red) and that of the star S2 (in blue) for comparison. The position of the black hole is represented by a red point, near the crossing of the two orbits, in (0,0). The dashed part of the orbit is an extrapolation. Right, the velocity of the gas cloud with respect to the black hole is plotted as a function of time. The gas cloud has reached its pericenter in 2013-4. From Stefan Gillessen. with thanks.

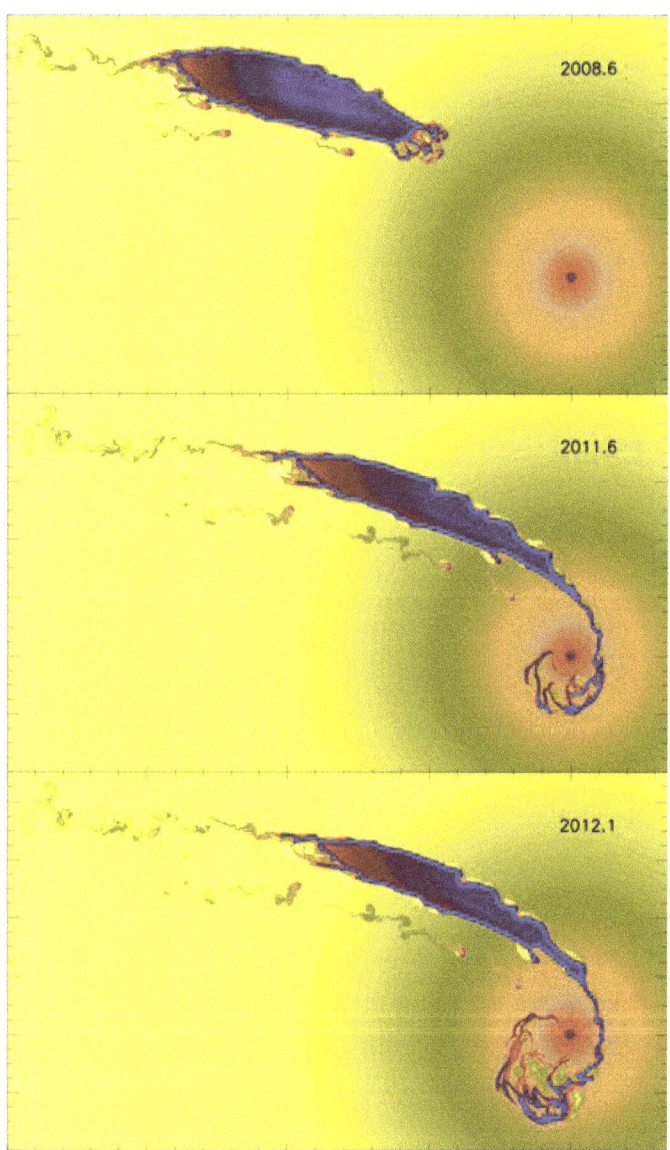

FIG. 4.10 – Simulations of a gas cloud on its orbit around the super-massive black hole of our Galaxy, arriving at pericenter in 2013-4. The date is indicated on each image, which dimensions correspond to 0.0235 × 0.0125 parsec. The cloud is stretched by tidal forces on its orbit, and part of the gas is leading in front of the cloud center on this orbit. From Burkert, A. et al. (2012) *Astrophysical Journal* 750, 58, with permission of the American Astronomical Society.

## 4.4 Conclusion

The center of our Galaxy is the place of unique dynamical events, whose study is very useful to better understand the active nuclei of galaxies. The stellar bulge is very modest, and more similar to what we call a pseudo-bulge, that is to say an intermediate component between bulge and disk. A classical bulge does not rotate, has a rather round shape and a typical light distribution in exp $(-r^{1/n})$, where $n = 4$. Conversely, disks rotate, are flat, and their light distribution is exponential ($n = 1$). The pseudo-bulge is somewhat flattened, rotates slightly (although more slowly than a disk) and its distribution of light corresponds to $n \sim 2$. The pseudo-bulges are formed from the disk stars, by resonance with the vertical bar. This suggests that the bar plays a crucial role in the evolution of the Milky Way, which has not undergone massive encounters with neighboring galaxies over billions of years. The gas then accumulates in a ring 200 pc in radius. Then it moves towards the center; strong non-circular motions, and different inclinations on the galactic plane, are the source of surprising morphologies difficult to elucidate because of projection effects. Some ionized gas falls onto the black hole whose mass is 4 million solar masses; This mass was determined by the patient study of proper motions of stars around the black hole. Short quasi-periodic flares (of period 20 minutes) help to trace the gas on the last stable orbit around the black hole. New observations at high spatial resolution with infrared interferometry will soon enable better knowledge of the super-massive black hole, its rotation, and how it was formed. It is a prototype helping to derive the properties of the black hole that is lurking in the core of every galaxy.

# Chapter 5
# Galactic dynamics

The Milky Way is a barred spiral galaxy; but as we are inside the disk that we see only by the edge, it is very difficult to have a clear and detailed picture of its spiral structure. It is only by comparison with external galaxies of the same type that we can deduce the most probable structures, aided by kinematic and photometric observations at different wavelengths. How do these structures form? What are their lifetimes and their role in the evolution of the Galaxy?

## 5.1 Dynamics of the barred spiral structure

The various tracers of the spiral structure bring different and complementary points of view, according to their nature. The gas component has some viscosity, implying a dissipative dynamics and a low velocity dispersion. This viscosity is due to collisions between interstellar clouds, collisions that dissipate their relative kinetic energy: the gas radiates this energy, often through shock waves, which are highly effective from that point of view. Stars are born by gravitational instability in dense molecular clouds; that is why the young stellar component, especially the OB-type massive stars that ionize the surrounding gas forming an HII region, has the same morphology and the same kinematics as the gas. The young and massive stars do indeed have a lifetime of about ten million years at most, which is very short with respect to the crossing time of a spiral arm at their velocity, for example.

However, the oldest stellar component has a very different dynamic. It is this component that contains the bulk of the mass; it consists of stars of the order of a solar mass, whose lifetime is several billion years. The number of stars in the Galaxy is so large (approximately 200 billion) that they produce a quite smoothed average gravitational potential, where the influence of individual stars is washed out. The ensemble of stars can be regarded as a collisionless medium, and therefore non-dissipative. The term "collision" here does not mean physical collision (the stars dimensions are so small relative to their average distance that they almost never collide), but should be extended to "gravitational encounters" during which stars could exchange energy. However, the time for the stars to exchange energy in pairs and settle in "thermal" equilibrium by equaling their kinetic energy is millions of times greater than the age of the Universe as we have seen in the previous chapter.

So, it is conceivable that the spiral structure of the gas and young star tracers is different from that of the old stars, as shown in Figure 5.1. The barred spiral structure is a density wave, which mainly develops in the most persistent stellar component, that is to say, the old stars, but with the help of gas. The theory of the formation of these density waves by gravitational instability of the galactic disk was developed in the 1960s by Lin, Shu and their collaborators, then confirmed by numerical simulations. As we saw in Chapter 3, the density waves explain why the majority of disk galaxies have spiral arms, while all the matter rotates differentially, deforming structures: the period of rotation of the gas and stars at the edge of the disks is of the order of a billion years, while in the center it is only about ten million years. Any material structure like a spiral arm, forming at one epoch should be wound at least 100 times during the life of a galaxy, and disappear through confusion. A density wave, instead, rotates at a constant angular velocity (the rotation period is therefore also constant) and can therefore persist without deformation, at least in the area of the disk where its existence is permitted.

Old stars therefore trace these two-arm density waves, relatively long-lived, that we see on Figure 5.1 at the right. They are not sensitive to small perturbations, which would be created for instance by the encounters between stars and giant molecular clouds. As for the gas, it follows the gravitational potential produced by these spiral arms, but it is much more disturbed than the old stars by various factors such as the formation of stars. The later may be contagious because the shock waves created by the winds of the young stars and the explosion of the most massive stars in supernovae stimulate subsequent star formation in molecular clouds: these processes generate random pieces of new arms or branching between the main arms. This explains the great complexity of the interstellar matter distribution, and therefore that of young stars and HII regions that they ionize. Some chaos is thus created in the disk structure, and higher harmonics of arm waves can develop: this is the origin of the four arm structure that was originally revealed by the HII regions in the Milky Way (see Figs. 1.6 and 3.26).

The density wave is not rotating at the same speed as the stars and gas in the galactic disk. So there are distances to the galactic center where resonances occur between the movements of matter and movement of the wave: see box E5.1. The most common resonance is the co-rotation, where the angular velocity of the spiral or barred wave is equal to that of the matter. Within this co-rotation radius (CR), matter rotates faster than the wave, and less rapidly outside the CR. Higher-order resonances also exist, where a rational combination of the rotation frequency and the oscillation frequency of the stars around their equilibrium point (the epicyclic frequency) is equal to the frequency of rotation of the bar. Because of this rational relation, the orbits of stars and interstellar matter near the galactocentric radius where such resonance occurs are closed (periodic) in the rest frame of the wave: the interaction between the wave and stars or gas can become very strong. The resonances of this type which are the most important in the galactic disk are the Lindblad resonances (box E5.2), where stars run two epicycles for each rotation completed in the spiral wave frame, either in the forward direction (inner resonance, in which the stars rotate faster than the wave) or in the retrograde direction (outer resonance, where the opposite is true).

Lindblad resonances and co-rotation play a major role in the development of density waves. These spiral waves can develop only between the two Lindblad resonances, inner and outer, and bars generally end shortly before the co-rotation. This is easy to understand, when we know that the major families of stellar orbits have their long axis parallel to the bar just inside the co-rotation ($\times 1$ orbits), while their major axis is perpendicular to the bar outside: see box E5.2. As the bar results from the arrangement of stellar orbits, the stars cannot sustain the bar beyond the co-rotation. The orientation of the orbits also changes from parallel to perpendicular at each resonance. The presence of the inner Lindblad resonance weakens the bar, and thus prepares the apparition by decoupling of a secondary bar, rotating faster, inside this inner resonance.

FIG. 5.1 – Two artist views of the Milky Way, which give an idea of what may be its structure. On the left, the gas component and young stars that form in molecular clouds (associations of O and B stars) and the HII regions ionized by these stars. These tracers show a 4-arm structure. On the right, the old stellar component, which represents the bulk of the disk mass, as would be seen in the near infrared (the scale is a little different). In this component, the Galaxy has two major spiral arms that may have some branchings, and weak harmonics of the main arms. The name of the main arms is indicated. Compare to the real images of external galaxies reproduced in Figures 1.7, 3.25, 5.3, 5.8 and 5.9. From R. Hurt, with thanks.

In the gravitational potential of the stellar bar, the gas tends to follow the same orbits as the stars, but its collisional and dissipative character imposes to the gas a different behavior. Clouds collide mainly when two streams intersect (parallel and perpendicular orbits for example), and all orientations then become possible by precession. The gas settles in a spiral structure, slightly out of phase with the stars, so that a torque is exerted between stars and the gas. An exchange of angular momentum occurs, which can vary depending on the position with respect to the resonances in the Galaxy. It can be shown that within the co-rotation the gas loses angular momentum, and gradually falls toward the inner Lindblad resonance. There it is in phase with the stars, at equilibrium. That is why the inner resonance is a place of gas accumulation, where new stars can form (Fig. 5.2). In the Milky Way, there is indeed a molecular gas ring, at a radius of about 200 pc, which corresponds to the inner Lindblad resonance of the bar.

The co-rotation of the bar would be about 4-5 kpc from the center, which corresponds to a pattern speed of the bar of 40 km s$^{-1}$ kpc$^{-1}$, or to a rotational period of 160 million years. As with the vast majority of spiral galaxies, the Milky Way has spiral arms in the "trailing" sense, that is to say that wrap in the direction opposite to the rotation (in Fig. 5.2, the Galaxy is supposedly viewed from the galactic North Pole and the sense of rotation is clockwise). This sense of winding for the arms is dictated by the minimization of energy: it is with this winding sense that the matter will gradually in-fall towards the center, where the potential energy is minimal. Outside the co-rotation radius, the matter rotates more slowly than the wave and enters into the spiral arms by their convex side. This is the case of the Sun, which in the arm reference frame rotates counterclockwise and will soon meet the harmonic arm Sagittarius-Carina, then the main arm Scutum-Crux-Centaurus, while receding from the Perseus arm (see Fig. 3.26).

If the torques due to the bar and spiral are driving the gas inside the co-rotation to the inner Lindblad resonance, it is the opposite that happens outside the co-rotation, where the gas is driven outwards to the outer Lindblad resonance (OLR). In external galaxies, it is easy to see the rings formed through gas that accumulates there (Fig. 5.3). In the Milky Way, this OLR ring would not be very far from the Sun; it depends on the rotation curve, which is not well known in these external parts. Some astronomers have identified kinematic disturbances on the stars observed in the solar neighborhood, which have been attributed to the presence of the outer resonance.

Galactic dynamics

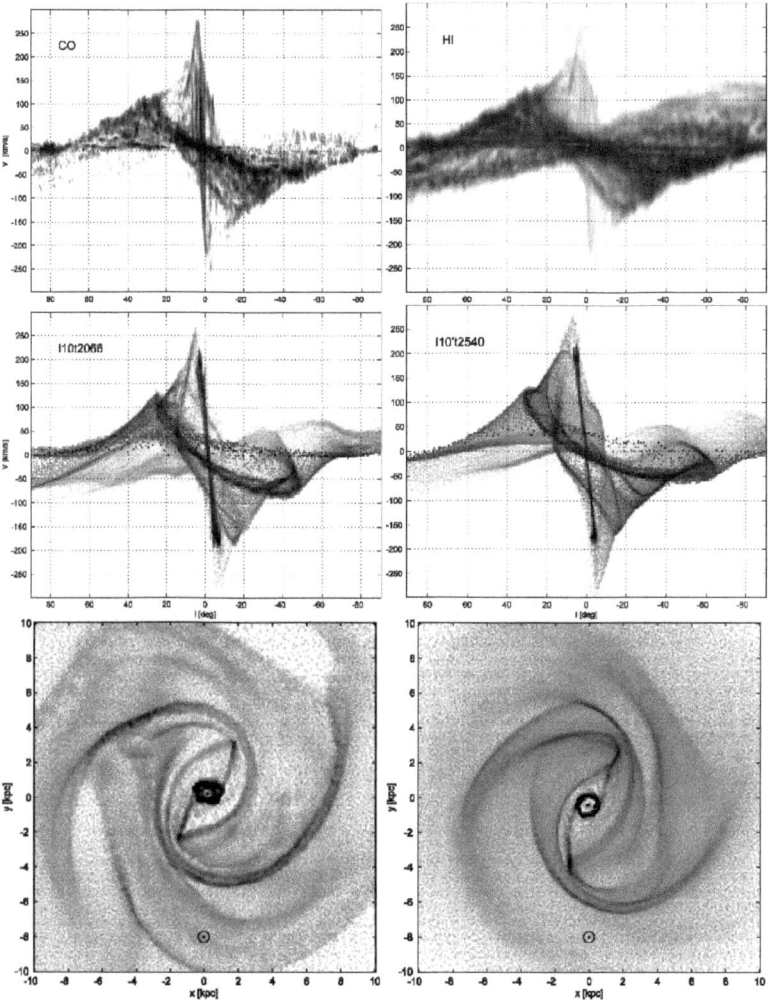

FIG. 5.2 – N-body simulations of stars and gas in a galactic disk, with the physical conditions of the Milky Way. The Figure corresponds to two of the best models and confronts them with the observations of the gas. Top left the longitude-velocity diagram observed for the CO emission (molecular gas), and top right for the HI emission in the 21 cm line (atomic gas). Middle: longitude-velocity diagram predicted by two possible models of the Milky Way. Bottom: projection viewed from the galactic North Pole of the interstellar gas distribution in both models. The Sun is at 8 kpc from the Galactic center, at the coordinates $(x, y) = (0, -8)$ kpc (symbol $\odot$). The co-rotation of the barred spiral wave is at $R(\text{CR}) = 4.5$ kpc. The matter and the wave rotate clockwise. Each of the models reproduces better some spiral arms and less some others, but both succeed to account for the global aspect of the observed Galaxy. From Fux, R. (1999) *Astronomy & Astrophysics* 345, 747-812, with permission of ESO.

FIG. 5.3 – The barred spiral galaxy NGC 1097, which could resemble our Galaxy. This is a 3-color image in the infrared taken by the IRAC camera on the Spitzer satellite: *blue* and *green* correspond to old stars observed at 3.6 and 4.5 microns, and *red* to dust whose emission dominates at 8 micron wavelength. The bar in this galaxy has produced 3 rings at resonances: a nuclear ring in the center at the inner Lindblad resonance, location of a starburst, a ring at co-rotation around the bar, and a ring at the outer Lindblad resonance, which is in formation (or rather a pseudo-ring, made of two spiral arms that begin to wind into a ring). The ensemble is disrupted by a small companion in the upper right, which is seen only in blue (old stars), and seems devoid of dust and gas. © CalTech/JPL.

Resonances between the motion of the stars and the motion of the bar also exist in the vertical direction (perpendicular to the plane of the Galaxy). By analogy with what happens with the radial oscillation (epicycle), it is easy to understand that an inner Lindblad resonance may exist perpendicular to the galactic plane, where the oscillation frequency perpendicular to the plane is equal to the rotation frequency of the bar: in these conditions, the star finds the bar in the same configuration each time it comes back to the disk plane. This resonance occurs in the central regions of the Galaxy, quite near the in-plane inner Lindblad resonance. The disturbance of the bar is then amplified for the resonant stars. The stars at this radius gain energy and are projected higher above the plane. The bar acquires the shape of a peanut. Examples are shown in Figure 5.4. The presence of this particular shape in our Galaxy is another confirmation of the presence of a strong bar.

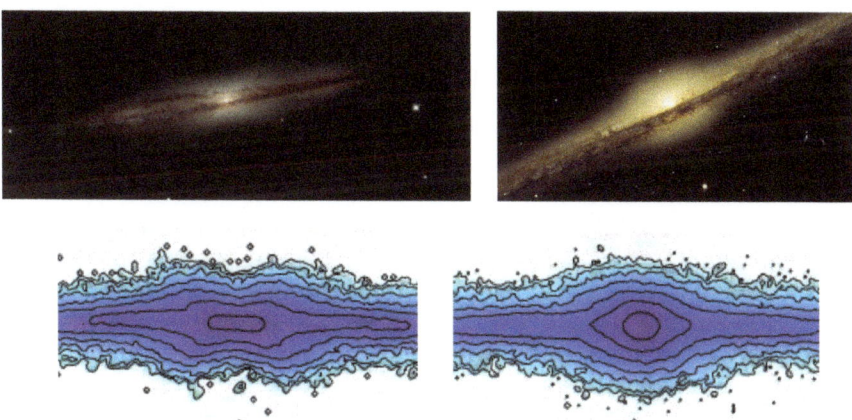

FIG. 5.4 – Top, two barred galaxies seen edge-on, showing a peanut shape bulge (Hickson 87-A, at left) or a box shape (NGC 4565 at right). The images come from the Hubble Space Telescope (© *Hubble Space Telescope Heritage*). Bottom, the results of N-body simulations show similar morphologies, when the bar is observed edge-on at 45° (left) or 10° (right) from the line of sight.

The consequences of the exchange of energy and angular momentum between matter (stars and gas) and the bar are all the more visible as the barred spiral wave is stronger and longer lasting. In the simplest case of a strong bar, there is only a single spiral wave with a single rotation period throughout the disk, the same as the bar period. However, computer simulations show that the disk dynamics can be more complex as it can develop several waves with different speeds, and that the interaction between gas and stars can lead to destruction of the bar.

## 5.2 Cycle of the bar evolution, migrations, multiple waves

How long will the bar in the Milky Way remain? When was it formed? The answer to these questions is not easy. It is not enough to determine the age of the stars that form the bar today, or even those that constitute the pseudo-bulge formed from the bar, to obtain the age of the bar. Because when the density wave develops by gravitational instability, it collects all the matter, gas and stars, regardless of their age. In a way, the bar does not have the age of its stars. In the pseudo-bulge we find mostly old stars, but also some young stars. Once the stars are propelled, through the vertical resonance, at a large height above the galactic plane, they will not lose their kinetic energy any longer. So it is possible to say that the bulge of our Galaxy accumulates all the stars brought there by the different processes encountered during the life of the Milky Way: accumulation due to the several bars that have succeeded, or even stars brought by a companion galaxy, which has lost some of its mass when flying near the Galaxy, or would have completely merged with it.

## 5.2.1 Destruction and re-formation of bars

When the mass of the disk contains more than 6% of gas, which is the case of the disk of the Galaxy, the angular momentum exchanges due to gravity torques can weaken the bar. Incidentally, the dark matter is not dominant in the central parts of the disk, and plays no significant role in this exchange. The torques are such that gas is driven inwards to the center and loses momentum. But the total angular momentum is conserved across the galaxy provided it is isolated; it is therefore an exchange between the gas and the stars forming the bar: gravity torques are reciprocal, because of the equality between action and reaction, which means that the gas exerts an opposite torque on the stars, which therefore will gain momentum. Now the bar wave is characterized by a negative momentum within the co-rotation radius, since it rotates slower than the matter. Giving angular momentum to the bar means weakening it and even destroying it if there is enough gas. In a way, it is a self-regulating process of the bar, which is shown in Figure 5.5.

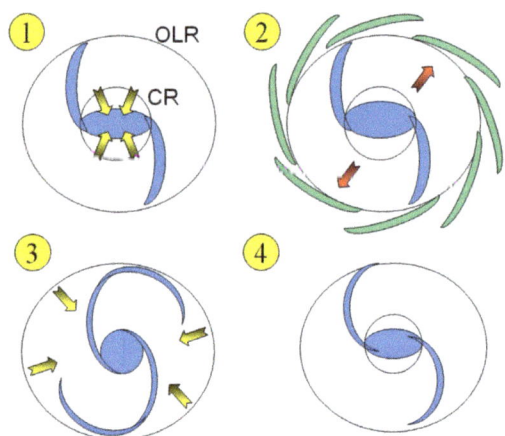

FIG. 5.5 – The evolution cycle of barred spiral waves. (1) In a first step, a wave grows by gravitational instability. The central bar exerts a torque on the gas, which follows the wave with a phase shift, due to its dissipative character. The gas inside the co-rotation (CR) is precipitated to the center, and accumulates on the inner Lindblad resonance. (2) Meanwhile, the gas outside the CR is driven outwards and accumulates at the outer resonance (OLR = Outer Lindblad Resonance). (3) The gas has now transferred all its angular momentum to the bar. The bar is a wave of negative momentum as it spins more slowly than the matter: it therefore weakens when accepting angular momentum. The galaxy's disk no longer having a strong bar, the gas that was forced to park outside the OLR can now enter. The average disk cools down (in average the stellar velocity dispersion decreases, since new stars form out of the cold gas) and becomes unstable. (4) Another barred wave can then be formed. As the mass was concentrated during this cycle, the new bar rotates faster than the disk and ends at its co-rotation.

*Galactic dynamics*                                                                 111

In numerical simulations, supported by observations that can estimate the magnitude of torques by measuring the phase shift between the stellar bar and the gas response, the in-fall of gas and the destruction of the bar can occur in a single dynamic time (or the characteristic time for a revolution), typically about 200 million years if the galaxy is rich in gas. But the cycle is slower if the galaxy is gas-poor, on the order of a billion years.

Once the bar is weakened or destroyed, the stellar disk has been "heated" by the passage of the wave, that is to say that the stellar velocity dispersion has increased. One has then to wait until the galaxy acquires a large amount of external gas from cosmic filaments, connecting to the rest of the Universe, for the disk to cool down and become gravitationally unstable again. Other barred spiral waves can then form. Figure 5.6 illustrates this process by numerical simulations taking into account (or not) external gas accretion.

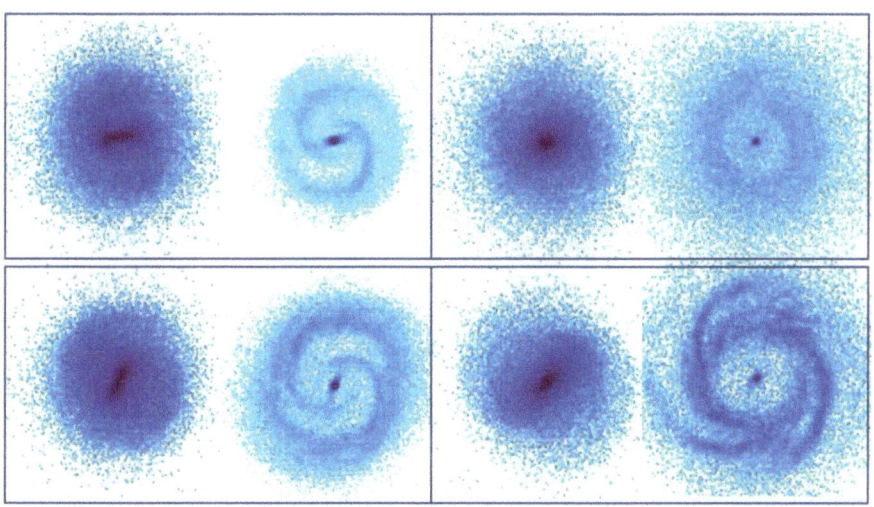

FIG. 5.6 – Simulations of the evolution of a galaxy similar to the Milky Way. Each of the four panels show, left, the distribution of the old stars, and, right, the gas distribution. The two left panels correspond to an epoch $T = 1500$ million years after the beginning of the simulation, and the right ones to $T = 8000$ million years. Top, there is no gas accretion by the galaxy, and the bar which developed at the beginning has disappeared at the end of the simulation. Bottom: with the same initial conditions, there is now external gas accretion by the galaxy. The bar persists until the end of the simulation.

## 5.2.2 Migrations

The barred spiral wave produces a large radial migration of matter (stars and gas) in the Galaxy. The stars we see today at some distance from the center stay there only transiently, and it is possible that the Sun was not born in its present distance from the galactic center, about 8 kpc, but in a more central area. It is important to take into account the effect of migrations in order to explain the abundance gradients of heavy elements according to the radius, or the stellar age gradients: this will be discussed in chapters 6 and 7.

These migrations are the result of a scattering of stars at the co-rotation resonance. This concerns mainly the stars that have little velocity dispersion, so that their orbit is nearly circular. In the rotating frame, it is possible at the co-rotation to change the sign of the angular momentum without virtually changing the energy, and it is very easy to switch from one side to the other of the circular orbit. There is exchange of angular momentum between the stars: the stars whose orbits have a smaller radius than the co-rotation will gain momentum, allowing those on the other side of the co-rotation radius to lose momentum and migrate toward the center. As the pattern speed of the wave varies over time, the radius of co-rotation varies, and the migration involves an increasing number of stars. The effect can also be amplified if several density waves exist simultaneously.

Figure 5.7 illustrates the characteristic effects of migration due to the barred spiral wave of the Milky Way. The numerical simulation shows that in 3 billion years, the radius of the disk of stars increases considerably because the stars in the center migrate to the outer parts. The changes in momentum occur primarily at resonances, but not only, because there are multiple waves with overlapping resonances. Not only stars and gas are spread in radius and form an exponential disk, but in this simulation the mean metallicity of stars rises up to the edge of the disk, which is paradoxical. If there were no migration, the abundance of heavy elements would decrease monotonically with the radius, as seen in Chapter 6. However, migration drives the old metal-rich stars that come from the center of the Galaxy outwards, which greatly affects the abundance gradient of heavy elements. It is striking that the birth radius of stars that are currently at a given galactocentric radius can take almost all possible values in the disk.

# Galactic dynamics

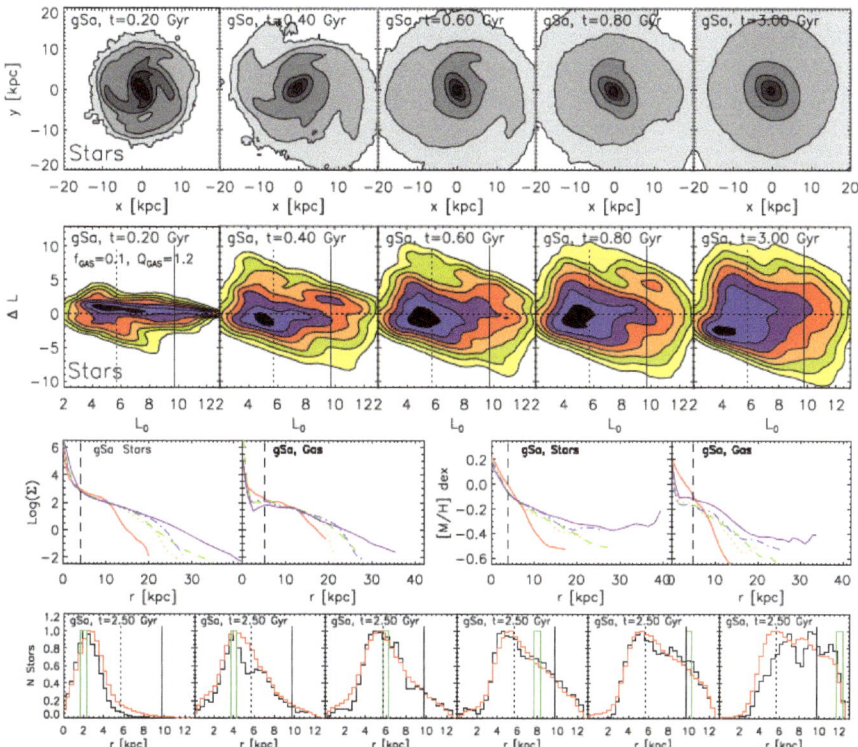

FIG. 5.7 – Migration of stars in a galactic disk. Top: Image of the stellar component in a simulation of a spiral galaxy without gas accretion, at times $T = 0.2, 0.4, 0.6, 0.8$ and 3 billion years; we see the gradual disappearance of the bar and of the spiral structure. Second line: For the same epochs, variation of angular momentum DL, normalized by the final rotation velocity $V_c$ of stars, as a function of radius; the co-rotation radius is indicated by a vertical dotted line, and the outer Lindblad resonance by a solid line. Third line: Radial distribution of the surface density of stars and gas (left) and the abundance of heavy elements in stars and gas (right) for the same epochs, starting from the red line at 0.2 billion years to the purple one at 3 billion years. The vertical dashed line indicates the initial radial scale. Fourth row: Distribution of the birth radius of the stars which are found at $T = 2.5$ billion years inside the ring shown in green (600pc in width). From Minchev, I. et al. (2011) *Astronomy & Astrophysics* 527, A147, with permission of ESO.

## 5.2.3 Secondary bar, multiple waves

When a galactic disk becomes unstable with respect to the spiral or bar density wave formation, the simplest case is the formation of a single wave, at a fixed pattern speed, between the inner and outer Lindblad resonances. Thanks to the self-gravity, this wave drags with it all the elongated stellar orbits, which then all precess at the rotation speed of the bar, whereas in the absence of the bar, the precession rate of each orbit would decrease with radius. The precession rate is a combination of the frequency of rotation and of the epicyclic frequency, both of which vary widely from center to edge of the Galaxy; but the combination varies much more slowly than those frequencies with radius; therefore the density waves can be maintained over time, and do not wind out and tangle quickly as material spiral arms would do. Yet the precession rate varies somewhat, more quickly where the mass distribution is more concentrated in the galaxy. As the barred spiral wave concentrates the mass, the natural evolution goes in the direction of an increasingly rapid variation of the precession rate, so that the self-gravity will eventually no longer be sufficient to gather all the stars in a single wave. There will be decoupling into multiple waves, rotating at different pattern speeds, slower at larger distance from the center.

One of the most common examples of these phenomena in external galaxies is the decoupling between the bar pattern speed and that of the spiral arms outside of the bar. For strong bars, arms always rotate at the same velocity, which is that of the bar to which they are attached. But in some galaxies, the arms are detached from the bar and can begin at any phase angle with respect to the bar as shown in Figure 5.8. Calculations show that this decoupling does not make the two waves completely independent. Conversely, the two waves exchange energy, by non-linear coupling, at a resonance they have in common. In general the bar ends at its co-rotation, which can then be the inner Lindblad resonance of the spiral. If the number of arms is two ($m = 2$ symmetry), the coupling with the $m = 2$ bar produces harmonics, which are $m = 0$ and $m = 4$: these harmonics, although weaker than the primary arms, are observed in numerical simulations of the phenomenon.

FIG. 5.8 – Images of the galaxies NGC 1073 (left) and NGC 3992 (right). They show a phase shift between their bar and their spiral arms, which do not begin at the extremities of the bar. It is possible that the Milky Way is similar to these galaxies. © Hubble Space Telescope Heritage.

Another frequently encountered example is the decoupling of a secondary bar inside the primary bar. The two bars are embedded and exchange energy, the co-rotation of the secondary bar being the inner Lindblad resonance of the primary bar, where the gas has accumulated to give a starburst in a first step. These secondary bars are small, 100 to 800 pc in radius, and are often buried in the dust, so that one sees them only in the near infrared images. An example is shown in Figure 5.9, which represents the barred galaxy NGC 1097.

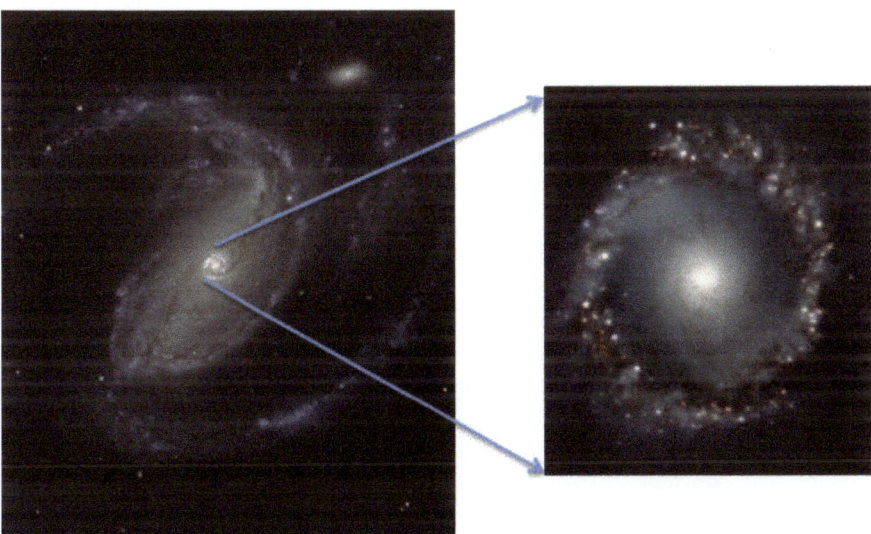

FIG. 5.9 – Images of the galaxy NGC 1097 (also shown in infrared in Fig. 5.3) obtained with the VLT telescope of ESO in Chile. The left image shows the primary bar, the spiral structure and a companion, at the upper right. The right image was obtained in the infrared with the NACO adaptive optics instrument on the VLT, and zooms into the nuclear ring, corresponding to the inner Lindblad resonance, location of a star formation, and the nuclear bar inside. It is quite possible that the Milky Way has such a ring and a similar nuclear bar. © ESO.

Note that it has been proposed that nuclear bars could help to feed the active galactic nuclei. The bars offer indeed the best way to bring the gas from the disk of the galaxy to the central black hole. The bar is a non-axisymmetric disturbance, which produces gravity torques acting on the angular momentum. However the primary bar could bring down the gas only from the co-rotation to the inner Lindblad resonance, where the gas is stalled in a ring since, as we have seen, gravity torques change sense inside the ring. Conversely, with a secondary bar that is inscribed in the ring, which is its co-rotation radius, the action may extend to the center. The flow of gas towards the galactic center thus goes in steps, and star formation bursts have time to occur during these stages. This explains why the activity of galactic nuclei, which takes place when some gas falls onto the black hole (see the previous chapter), is often, but not always, accompanied by an active nuclear star formation.

The lifespan of secondary bars is shorter than that of the primary bars, at least in the numerical simulations. They persist for several rotations, but the dynamic time is much shorter at the center since the rotation is faster. This agrees with the observational statistics of the nested bar phenomenon: among the strongly barred galaxies, only about 30% have nuclear bars.

Our galaxy has no nuclear activity, and goes through a quiet period as we saw in the previous chapter. Yet the star counts in the near infrared suggest the presence of a nuclear bar. As the ring at 200 pc radius is rich in molecular gas, it is likely that in a dynamic time (about 10 million years at these small radii), the gas will spiral in toward the center, and feed the black hole. Very near the nucleus, at 10 pc in radius, other dynamical processes should take over, as the dynamical friction that would slow down the giant molecular clouds amidst the bulge stars, or asymmetries off-centerings, etc. Only a cascade of such mechanisms can explain the efficiency of nucleus feeding, and these successive mechanisms occur at shorter and shorter characteristic time scales, as it approaches the nucleus.

The comparison of the dynamics of external galaxies with that of the Milky Way helps us to better understand what is happening around us. We can observe our Galaxy in great detail, but unfortunately only in the plane of the disk. We now know, thanks to simulations and comparisons with the external spiral galaxies, that the structures which appear robust and perennial in the Galaxy are actually relatively transient and constantly changing.

Even more violent changes can occur, caused by the interaction between galaxies, and this will be the object of Chapter 8.

### E5.1. Stellar orbits, epicyclic approximation

Each star in the disk of the Galaxy has to zeroth order a rotational motion around the center, with the angular velocity W, then to the first order of perturbations, a small oscillating motions around this circular orbit, which can be described by an epicycle: This is the so-called first order "epicyclic theory". The term epicycle is borrowed from ancient Greece, where the movement of the planets was described by superposing to a circular motion another motion on a circle, the epicycle. But here the epicycle usually has a shape close to that of an ellipse.

The nature of the orbits sheds light on the collective behavior of stars in the Galaxy: indeed families of periodic orbits exist in the rotating frame of the density wave (bar or spiral). These periodic orbits trap around them the majority of the other orbits as attractors. A small part of the orbits is not trapped and remains chaotic in nature.

To calculate the epicyclic frequency, let us represent the Milky Way as a flattened disk, axisymmetric, where the gravitational potential is $U(r, z)$, and let us write the equations of motion in the Galilean frame of cylindrical coordinates $(r, \theta, z)$. We consider nearly circular orbits, which depart by small deviations $x$, $y$ and $z$ respectively in directions radial, tangential and perpendicular to the plane. At zeroth order, the coordinates of the star are $(R, \theta = \omega t, 0)$, where w is the angular velocity. At the first order, this yields:

$$r = R + x$$
$$\theta = \omega t + y$$

From the dynamics principles, the angular velocity is such that:

$$\omega^2 = 1/R \; \partial U/\partial r \; (R,0)$$

Writing the equations of motion in polar coordinates, and developing to the first order of perturbations, one is led to develop in Taylor series the derivative $\partial U/\partial r$ of the potential $U(r,z)$, which is the force applying on the star:

$$\partial U/\partial r = \partial U/\partial r \; (R,0) + x \; \partial^2 U/\partial r^2 \,(R,0) + z \; \partial^2 U/\partial r \partial z \; (R,0)$$

The last term is zero, given the plane symmetry: $\partial U/\partial z(R,0) = 0$.

The first order expansion amounts to considering only the quadratic potential around the equilibrium position: small movements are then harmonic oscillators. Thus we find the equations of motion $x$ and $y$ in the plane of the disk relative to the circular motion:

$$d^2 x/dt^2 = -\kappa^2 x$$
$$dy/dt = -2\omega \; x/R$$

The frequency of these harmonic oscillations, $\kappa$, is the epicyclic frequency such as:

$$\kappa^2 = \partial^2 U/\partial r^2 (R,0) + 3\omega^2 = R d\omega^2/dR + 4\omega^2$$

In the same way, $z$ oscillates with the frequency $v_z$, such as

$$v_z^2 = \partial^2 U/\partial z^2 \; (R,0)$$

The orbit of the star in the disk is a combination of a circle and a epicycle, which is a small ellipse in the rotating frame of the wave: this is usually a rosette, as is shown in Figure E5.1. This epicycle is run in the opposite direction to that of rotation, and its axis ratio is $\kappa/2\omega$. We see in the equation giving the epicyclic frequency that it is of the same order of magnitude as $\omega$ but generally higher. Specifically, in the inner regions of the Galaxy where the rotation curve rises almost linearly, we have $\kappa \sim 2\omega$, then in the outer parts, when the rotation curve is flat, we have $\kappa \sim \sqrt{2}\;\omega$. As the number of epicycles per revolution is $\kappa/\omega$, we see that we can have an orbit that closes as an ellipse when $\kappa = 2\omega$.

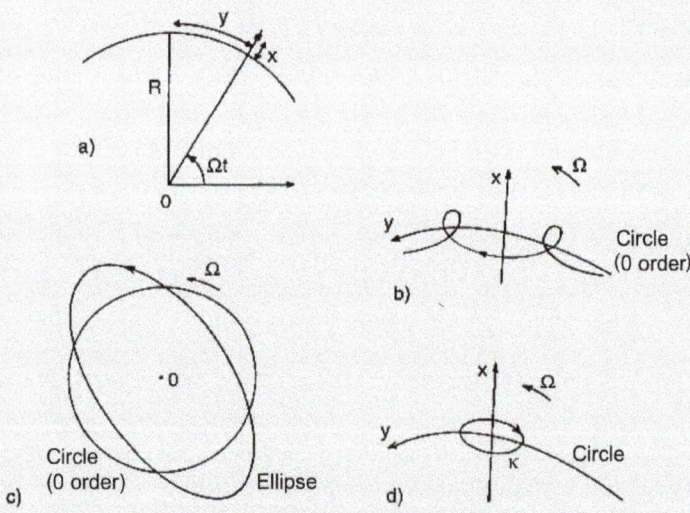

FIG. E5.1 – Orbits of stars in a disk according to the first order epicyclic approximation: a) definition of small displacements $x$ and $y$; b) any orbit (rosette), run in the opposite direction to the general rotation; c) particular case of a closed orbit where $\kappa = 2\omega$, corresponding to the central rigid rotation, or to a resonant orbit in the frame rotating at the speed $\omega - \kappa/2$; d) orbit of a star at the co-rotation resonance, in the rotating frame : the star does not make any revolution, but runs over only one epicycle.

## E5.2. Lindblad resonances

Consider the orbit of a star in the frame rotating with the density wave (bar or spiral). To better understand the interactions between the wave and stars, we must be placed in that reference frame, which rotates with the wave at the frequency $\Omega_p$. In this reference frame, the rotation frequency of the stars is $\Omega - \Omega_p$. The orbit of the star is periodic, closing in on itself, if there is a rational relationship between the frequency of rotation and the epicyclic frequency. The most frequent ratio actually occurring in galaxies is the ratio 2, i.e.

$$\Omega - \Omega_p = \kappa/2$$

We call this region the inner Lindblad resonance; it occurs within the co-rotation. When this ratio occurs with the opposite sign, it is on the outside of the co-rotation, and

$$\Omega - \Omega_p = -\kappa/2 ,$$

which defines the outer Lindblad resonance.

Figure E5.2 plots on the rotation curve of the Milky Way of Figure 3.30, the computed frequencies $\Omega$, $\Omega - \kappa/2$, and $\Omega + \kappa/2$ and shows how a barred spiral wave rotating at the pattern speed $\Omega_p = 38$ km/s/kpc could represent the observational data, the co-rotation lying between 5 and 6 kpc, and the outer resonance being just outside the solar radius. There would be two inner resonances, which clearly favors the decoupling of a nuclear secondary bar.

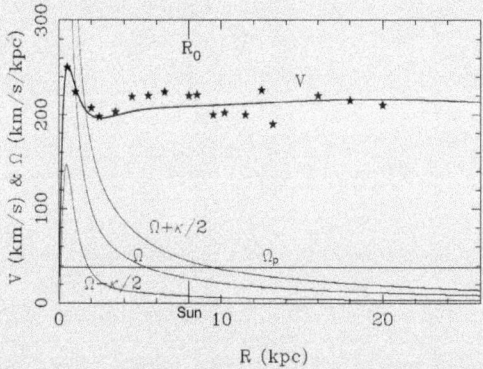

FIG. E5.2 – Rotation curve of the Milky Way (see also Fig. 3.27), where the characteristic frequencies $\Omega$, $\Omega - \kappa/2$, $\Omega + \kappa/2$ are computed, in order to determine the position of resonances with the density wave at a pattern speed $\Omega_p$.

In a barred spiral galaxy, within the co-rotation, there are two main families of periodic orbits, parallel (×1) and perpendicular (×2) to the bar (Fig. E5.3). The later exist when there are one or two inner Lindblad resonances.

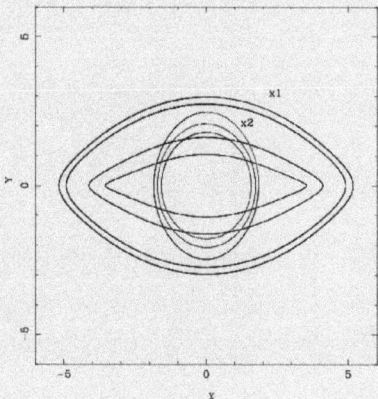

FIG. E5.3 – Shape of the main families of orbits in the potential of the bar (horizontal in the figure), inside its co-rotation. Closed orbits parallel to the bar (×1) and perpendicular (×2) exist, of which some examples are drawn here.

# Chapter 6
# The chemical evolution of the Galaxy

Our Galaxy, like all other galaxies, is continually evolving over time: stars are born from interstellar matter and eject some of their material into the interstellar medium during their life and at their death, from which other generations of stars are formed. In addition, our Galaxy exchanges matter with the intergalactic medium. We will deal with this evolution in this chapter, focusing on aspects related to the abundance of heavy elements in the gas and stars. The following chapter will examine in more detail the possible mechanisms for this evolution.

At first, about 13 billion years ago, the Galaxy was made entirely of gas and dark matter, and the first stars began to form. From this first generation, the only stars that remain today are those with a mass smaller than 0.9 $M_\odot$. The more massive stars have disappeared, having returned part of their material to the interstellar medium while the rest formed a compact star: white dwarf, neutron star or black hole, whose life is infinite or extremely long, except in some particular cases. Among the stars born later, those that have reached us have a higher limiting mass if they were born more recently. A growing part of the mass of the Galaxy is blocked over time in low-mass stars and compact objects. Consequently, the mass fraction contained in the interstellar medium can only decrease over time. However, fresh interstellar medium and stars from outside were, and are still captured by the Galaxy. Conversely, the Galaxy lose interstellar medium by emitting a wind, but it cannot lose many stars because of its large mass that keeps them captive. All this can be summarized by the diagram of Figure 6.1.

## 6.1 The formation of the Galaxy

This will be covered in the next chapter, but it seems useful to say a few words by way of introduction here.

Currently the most popular scenario among astronomers for the formation of galaxies (which does not mean that it is the right one because it has problems!) assumes that small condensations (mini-halos) of dark matter, whose mass was about $10^7$ $M_\odot$, were formed from primordial density fluctuations. They each caught some of the gas that filled the universe, then

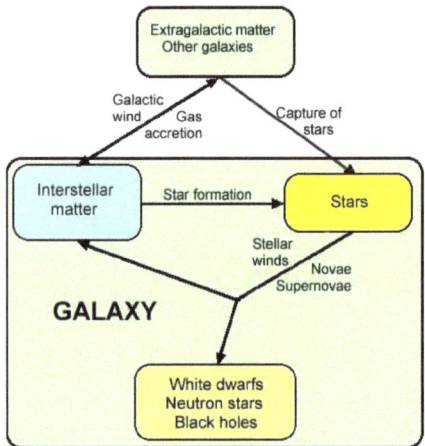

Fig. 6.1 – Diagram of the evolution of the Galaxy.

gathered gradually in proto-galaxies: numerical simulations show that half of these mini-halos had merged about 3 billion years after the Big Bang, 10.7 billion years before today. These proto-galaxies, dominated in mass by their dark matter halo, then gathered in clusters and other large structures; but some, like ours, have remained isolated in small groups of galaxies. Mini-halos left behind surrounded them, and formed dwarf satellite galaxies (but simulations predict too much of them). Galaxies observed at very large distances in the Universe, thus in the early stages of their evolution, are rather shapeless but already produce lots of stars: 2 or 3 billion years after the Big Bang, the star formation rate in the Universe averaged at least ten times more than now. Was this the case of our Galaxy?

The formation of the disk of a galaxy like ours is understandable if the protogalactic cloud was rotating, which should be a very common case. The centrifugal force prevents contraction perpendicular to the axis of rotation, while only the pressure of the gas is opposed to the contraction parallel to this axis. If the heat generated by the contraction of this gas can be radiated efficiently, the galaxy takes the form of a rotating disk, still contained in a halo of dark matter. Thereafter, gravitational instabilities, spirals and bars, which develop in the disk, redistribute the angular momentum between its different parts: we see that most of the angular momentum is presently in the outer parts of the disc, while the internal parts, including the central bulge, are slowly rotating. Meanwhile, plenty of stars are forming in the protogalaxy, and supernovae affect the evolution of gas.

The power of today's computers makes it possible to simulate these processes. The result is that one of the first stellar components to form, less than 4 billion years after the Big Bang, is a central bulge, where star formation is very active but stops quickly. In addition, a small number of stars are also born in the halo. The disk is formed later, 4 to 5 billion years after the Big Bang. All the stars formed after this period are in the disk, and some in the bulge. The first stars of the disk appear to occupy a greater thickness than the following ones, suggesting a possibility for the formation of the thick disk. However, these simulations currently have limitations because they often produce too massive bulges compared to those observed in real galaxies.

Numerical simulations were also carried out without dark matter, where the Newtonian law of gravity is modified at very large distances (the MOND models, see the end of Chapter 3). The results are not essentially different as regards the dynamics of galaxies. Much remains to be learnt, however, regarding their formation, but it seems that the coalescence between the initial fragments of material is rather easier in these models than in models with dark matter.

## 6.2 The production of elements in stars

At the time of formation of the Galaxy, the Universe contained only hydrogen and its isotope deuterium, helium and its isotope $^3$He, and a small amount of lithium 7, produced in the Big Bang. It perhaps also contained a small amount of heavier elements synthesized in the first stars (Population III), which are massive stars that have disappeared today. If we denote the *metallicity* by $Z$, that is to say the ratio of the mass of elements heavier than hydrogen and helium to that of hydrogen, it did not exceed $Z = 0.001$ in the interstellar matter after this nucleosynthesis by Population III stars, may be even less.

Most of the heavy elements have been synthesized in stars after the formation of the Galaxy. Here is the list of the main processes, with the sites where they occur in italics:

- Hydrogen fusion into helium, by the proton-proton reactions or the CNO cycle: *stars on the main sequence*;

- Combustion of helium to produce the so-called α elements ($^{12}$C, $^{16}$O, $^{20}$Ne, $^{24}$Mg, $^{28}$Si, $^{32}$S); auxiliary reactions produce other isotopes and other elements such as Na or Ca from these main isotopes and the by-products of the CNO cycle: *red giant stars and stars on the asymptotic giant branch;*

- Explosive combustion producing various elements up to iron group included: *novae, X-ray bursts, supernovae of all types*;

- Nucleosynthesis beyond iron by the s-process, which involves slow neutrons: *stars of the asymptotic branch;*

- Nucleosynthesis beyond iron by the r-process involving fast neutrons: *Type Ia supernovae*;

- Nucleosynthesis beyond iron by the p-process that produces neutron-poor nuclei: *supernovae*?

One generally expresses the abundance (in mass) of any element X relative to hydrogen H by reference to its abundance in the Sun, in the form:

$$[X/H] = \log(X/H) - \log(X/H)_\odot. \tag{6.1}$$

Similarly, an abundance ratio such as that of oxygen to iron is expressed as:

$$[O/Fe] = \log(O/Fe) - \log(O/Fe)_\odot. \tag{6.2}$$

Figure 6.2 shows the contribution of stars of different masses to the enrichment of the interstellar medium by the winds they emit during their lives, or by their final explosion in the case of stars more massive than about 8 $M_\odot$.

One reason for the difference that we see in Figure 6.2 between the production of heavy elements by stars of low metallicity and by those of high metallicity is that some new heavy elements are synthesized from other pre-existing heavy elements in the star. These new elements are called *secondary elements*, as opposed to the *primary elements* that can be formed from hydrogen and helium. $^{16}O$ and the α elements are the main primary elements, while most other elements are at least partly secondary, which greatly complicates the study of the chemical evolution of galaxies. For example, $^{14}N$ is generally considered a secondary element: indeed, the CNO cycle and side reactions synthesize $^{14}N$ from $^{16}O$, and also some $^{12}C$, $^{13}C$ and $^{22}Ne$, etc. But $^{14}N$ can also be primary. $^{12}C$, which is largely a primary α element because it is formed directly from $^{4}He$, can also be partly secondary.

The abundance of a secondary element produced by a star should be proportional to the initial abundance of the primary element from which it came, so this proportionality should be reflected in the chemical evolution of galaxies. That is pretty much the case for nitrogen with respect to oxygen, at least where the interstellar gas has a relatively high metallicity; but at low interstellar metallicity, the abundance of nitrogen seems independent of that of oxygen, which shows that there is also some production of primary nitrogen, probably from $^{12}C$, by stars of large and medium mass. This production dominates the ratio [N/O] at low metallicity because there is not yet much oxygen. Conversely, it is hidden by the secondary production when there is plenty of oxygen.

Figure 6.2 gives an incomplete picture of nucleosynthesis, for it shows only the contribution of isolated stars, and not of the very close double stars that are the novae (whose contribution is however rather negligible), and overall the Type Ia supernovae (SN Ia). Unlike type Ib or II supernovae which are massive stars, the SN Ia are initially a close pair of stars, the most compact component of which is a white dwarf composed of degenerate matter whose

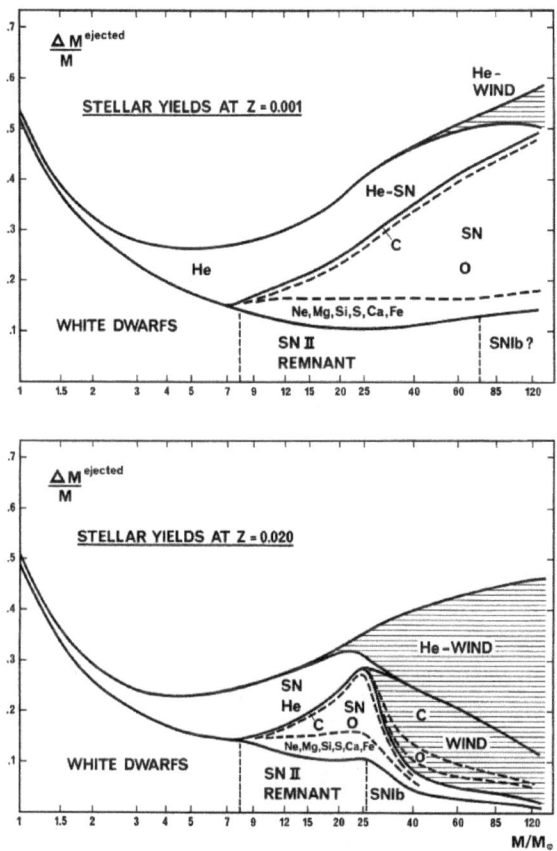

FIG. 6.2 – Production of heavy elements ejected into the interstellar medium by stars of different masses and initial metallicity $Z = 0.001$ (top) and $Z = 0.02$ (approximately the current metallicity of the interstellar medium near the Sun, bottom). The fraction of the initial mass blocked in compact objects after the death of the star is given by the lower curve. The ejected fraction of the initial mass of the star is above this curve, and the part that enriches the interstellar helium and heavier elements is between this curve and the upper curve. Above the upper curve, the ejected mass has the initial composition of the star and therefore does not participate in the enrichment. Note that massive stars contribute by their wind at high metallicity, but very little for stars with low metallicity. For very large masses, the stars could disappear entirely into a black hole when they explode as supernovae, in which case their contribution to the production of heavy elements would be zero. As seen in the figure, the effect would be especially important at low metallicity. From Maeder, A. (1992), *Astronomy & Astrophysics* 264, 105-120, with permission of ESO.

nuclei are mainly those of $^{12}C$ and of $^{16}O$. The standard scenario assumes that this white dwarf accretes matter from the other star. The mass of the accreted material compresses the white dwarf whose radius decreases from

10,000 km to 3,000 km, while its temperature increases. When the total mass of the white dwarf mass reaches the so-called Chandrasekhar mass limit, about 1.2 $M_\odot$, it can no longer be supported by the pressure of electrons in the degenerate matter and collapses, causing a sudden increase of temperature. When this temperature reaches $4 \times 10^8$ K, the fusion of $^{12}$C into $^{24}$Mg occurs, releasing a lot of energy. However, as the material is degenerate, the pressure does not increase and the material does not expand, thus cannot cool down: the nuclear reaction becomes explosive: in less than a second, the successive available fuels ($^{12}$C, $^{16}$O, up to $^{28}$Si) burn to form nuclei of the iron group. The SN Ia, which are slightly less numerous than the massive supernovae in our Galaxy (in total a supernova explodes every 50 years on average) are very important sources of iron group elements, the most abundant of which being obviously $^{56}$Fe.

The contribution of the massive supernovae (type II, Ib and Ic, which are also called core-collapse supernovae) to the nucleosynthesis of iron is rather uncertain, besides which, the SN Ia probably dominate the production. This happens quite late in the evolution of a generation of stars because it must first form white dwarfs, which are the result of the evolution of stars with masses below 8 $M_\odot$. Most of these stars have a mass much lower than this limit and need several billion years to produce white dwarfs. Figure 6.3 shows an example of the possible evolution of the appearance over time of SN Ia and of massive supernovae.

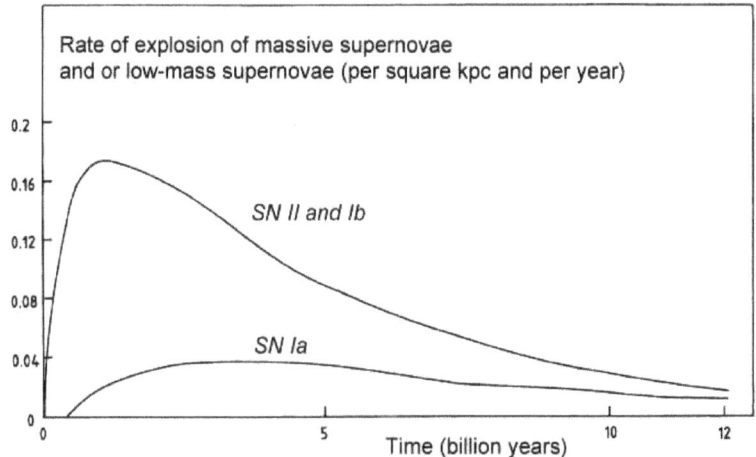

FIG. 6.3 – An example of the evolution over time of the rate of occurrence of massive supernovae (SN II, SN Ib) and of low-mass supernovae (SN Ia) in the Galaxy. The calculation corresponds to a specific model and is presented for illustrative purposes only. We observe the lateness of SN Ia compared to massive SN. From Matteucci, F. & Greggio, L. (1986), *Astronomy & Astrophysics* 154, 279-287, with permission of ESO.

*The chemical evolution of the Galaxy* 127

We can therefore expect that the enrichment in iron occurs well after that in α elements, which is mainly due to the stars of mass greater than 8 $M_\odot$, whose life is shorter than 20 million years (see Fig. 6.2 and Appendix 1). So, if an outbreak of star formation occurred very quickly at the beginning of the Universe, the second-generation stars were then formed from a gaseous medium enriched in α elements, but which has not had time to be enriched in iron. These stars will keep this signature if the star formation then stopped. But if the star formation is continuous, the abundance of iron will rise gradually, and the abundance ratio [α/Fe] between α elements and iron will decrease. The change in the relative abundance of α elements relative to iron thus offers a very important diagnosis for the history of star formation, and allows us to test various scenarios. We can empirically estimate the relative contribution of massive stars and of the SN Ia to the enrichment in iron by considering the logarithmic abundance ratio [O/Fe]: it is about 0.6 for the halo stars whose heavy elements were formed exclusively by massive stars, and falls to –0.1 for young stars, for which the contribution of SN Ia to the iron enrichment was important.

## 6.3 Modeling the chemical evolution

The modeling of the chemical evolution of galaxies is complex because it depends on many factors, which are likely to vary over time:

- The lifetime of stars of different masses;
- The mass distribution of stars at the time of their formation, which is the initial mass function (IMF);
- The star formation rate (SFR);
- The production of the different elements by the stars (yield);
- The ejection mechanisms of these elements (stellar winds, novae, supernovae);
- Their mixing with the interstellar gas;
- The interaction with the environment (gas accretion or ejection of interstellar medium by the galaxy);
- Last but not least, the dynamic mixing and gas migration and stars due to the gravitational action of density waves.

In practice, it is necessary to simplify the problem by assuming that some parameters are constant (that is what is usually done for the initial mass function in the absence of evidence for its possible variations), or by imposing their variation more or less arbitrarily. We will not write in detail the differential equations that govern the chemical evolution, equations that must be solved numerically. We will give only the very simple result obtained by neglecting the life of the stars with respect to the time scales of evolution, which are billions of years. This simplifying assumption, called the *instantaneous recycling approximation*, is approximately valid for the evolution of the interstellar material in general and for those of the primary elements that are mainly produced in high-mass stars, in practice the α elements and oxygen. If moreover the evolution is in a "closed box" without exchange of matter with the outside, we find:

$$M_g/M = \exp(-Z/y_z) \;, \tag{6.3}$$

where $M$ is the total mass, $M_g$ the mass of the interstellar gas, $Z$ the abundance in mass of the considered element (e.g. oxygen) compared to hydrogen, and the yield $y_z$, which is the mass ratio of this newly synthesized element to the total mass remaining in the stars and compact remnants of the corresponding stellar generation after it has been ejected. Equation (6.3) quantitatively expresses the fact that when the fraction of mass in the form of gas decreases, the gas and newly formed stars are enriched in heavy elements.

We can calculate the net yield $y_O$ for oxygen by assuming an initial mass function, and integrating the production of oxygen over this mass function as shown in Figure 6.2. However, $y_O$ is not constant over the chemical evolution: it changes gradually as the metallicity of stars increases. It also depends on what happens when massive stars explode as supernovae: if they disappear entirely into a black hole, which could occur for very large masses, $y_O$ would obviously be lower since these stars do not reject oxygen. Thus, there is a great uncertainty about the value of $y_O$ to adopt in equation (6.3). However, we can try to determine it empirically by applying this equation to external galaxies. Figure 6.4 shows the result. An average value $y_O = 0.0027$, corresponding to a total yield of heavy elements $y_z$ around 0.015, looks appropriate for spiral galaxies like ours. However $y_O$ appears to be 1.5 to 4 times smaller for irregular galaxies, which generally have a lower metallicity. This is paradoxical at first glance, as we expect $y_O$ to be greater at low metallicity, at least if the minimum initial mass of stars that disappear completely as black holes is greater than 25 M$_\odot$ (see Fig. 6.2). If the minimum initial mass for which the stars disappear into a black hole is significantly less than 25 M$_\odot$ $y_O$ is smaller at low metallicity, which may account for the observations. To complicate matters, it is possible that the gas in the spiral galaxies is diluted by accretion of intergalactic gas, in which case $y_O$ would be even greater than 0.0027, while irregular galaxies cannot accrete gas because of their low mass. Therefore we cannot conclude anything quantitatively from Figure 6.4. This illustrates the difficulties that we have in interpreting the relationship abundance/gas mass fraction in galaxies.

We have hitherto considered the metallicity of galaxies as a whole. What if abundance gradients are considered within a galaxy like ours? Since the gas/stars ratio increases with radius in the disk of the Galaxy, it is expected

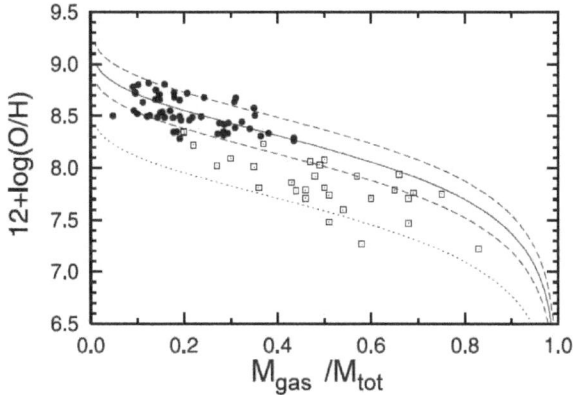

FIG. 6.4 – Abundance of oxygen O/H as a function of the $M_g/M$ ratio for spiral galaxies (black circles) and for irregular galaxies (empty squares). The solid line represents the prediction of the instantaneous recycling model in the case of closed-box evolution, with a yield of oxygen $y_O \approx 0.0027$; the curves in broken lines correspond to values of $y_O$ 1.5 times larger or smaller. The dotted curve is for a yield four times smaller. From Pilyugin, L. S. et al. (2004) *Astronomy & Astrophysics* 425, 849-869, with permission of ESO.

that the evolution of the outer regions is less advanced than that of the inner regions, and that their metallicity will be lower. This is actually the case: the abundance of oxygen, as measured in HII regions, decreases in our Galaxy when we go to the outside, while there is more and more mass in the form of gas (Fig. 6.5). We should however be wary of an immediate interpretation of this relationship, as we will see later.

To study the abundance gradient in the galactic disk, we can also consider the chemical composition of young stars: it is in principle similar to that of the gas that gave birth to them, and has the advantage of being measured further away from the galactic center. The abundance gradient of heavy elements observed in cepheids, which are young, massive stars, confirms what is presented in Figure 6.5. The behavior of the abundance of iron in open clusters, however, looks different (Fig. 6.6): this abundance is approximately uniform for galactocentric radii larger than 8.5 kpc. The difference with the result for young populations could come from an age effect, because these clusters are all old. A similar effect was observed in the spiral galaxy M 83, but this time concerning the abundance of oxygen in HII regions, as is so in the gas. We probably should, in both cases, invoke dynamical effects: mixing and migration of gas and stars. Another problem is raised by the presence of relatively young red giants in the solar neighborhood that are richer in a elements than the Sun: they probably migrated, like the Sun itself, from inner regions of the Galaxy, or were born from "pockets" of high-metallicity gas. In any case, we can only note that the chemical evolution of the Galaxy, which was believed to be simple only two decades ago, is blurred by dynamical processes that make its study complex. We have already seen another example of this complexity in Chapter 5 (see Fig. 5.7).

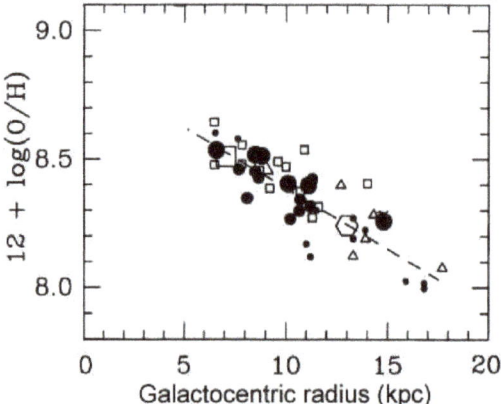

FIG. 6.5 – The gradient of the abundance of oxygen in the Galaxy, measured in the HII regions. The symbols correspond to observations made by various authors and reduced in a uniform way; their size is greater if the result is safer. From Deharveng, L. et al. (2000) *Monthly Notices of the Royal Astronomical Society* 311, 329-345, Wiley, with permission of the publisher.

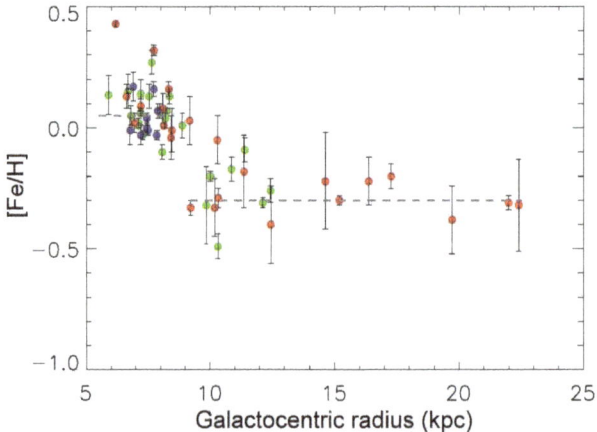

FIG. 6.6 – Abundance of iron in open clusters as a function of galactocentric radius. The radius at the Sun is taken here as 7.5 kpc. The abundance is relative to the solar one: [Fe/H] = log (Fe/H) – log(Fe/H)$_\odot$. The color indicates the age of the cluster: blue, clusters younger than 200 million years; green, from 200 to 1200 million years; red, clusters older than 1200 million years. The straight dashed line indicates the average value of [Fe/H] beyond 8.5 kpc. From Lépine, J. et al. (2011) *Monthly Notices of the Royal Astronomical Society* 417, 698-708, Wiley, with permission of the publisher.

## 6.4 The chemical evolution of the halo and the bulge

Despite all of the uncertainties described in the previous section, it remains that the abundances of elements in stars are markers of their origin and provide important information, complementary to those provided by their kinematics and dynamics. Moreover, there are two cases where the evolution is relatively simple: the halo and central bulge of the Galaxy.

As we have seen, the galactic halo has a complex stellar composition. A portion results from the capture of dwarf spheroidal galaxies and forms identifiable streams. However, the globular clusters of the halo, and the number of stars they have lost over time and that are scattered everywhere, have a very low metallicity and have certainly been among the first elements of the Galaxy to be formed, some 12 to 13.5 billion years ago. None of these clusters or isolated stars are completely devoid of heavy elements, because the initial gas in the Galaxy was enriched by Population III stars. The age of the halo stars is still too poorly determined to enable us to determine an age-metallicity relation. However, the younger globular clusters are richer in heavy elements. The question is, have they been formed *in situ* or were they captured from dwarf spheroidal galaxies? It is not possible today to answer this question.

Figure 6.7 shows the relation of metallicity with age for different stellar components of the Galaxy.

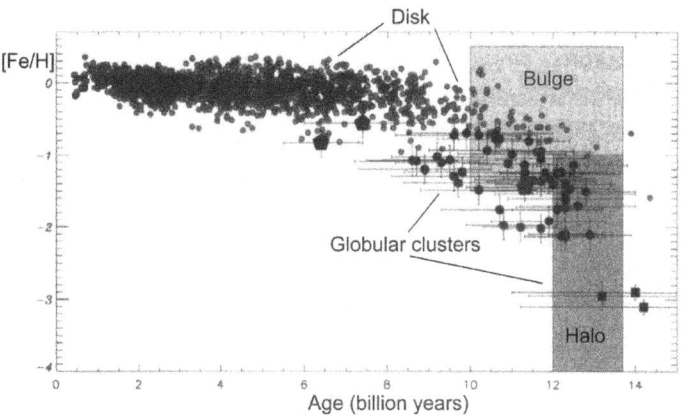

FIG. 6.7 – The relationship between age and iron abundance for stars of the disk, of the halo and of the bulge of the Galaxy. The abundance is relative to the solar one: [Fe/H] = log (Fe/H) − log(Fe/H)$_\odot$. The small points correspond to the disk stars, and larger circles with error bars to globular clusters. The three squares at the bottom right corresponding to halo stars are dated by cosmochronology, a technique that involves using the abundances of the detected radioactive elements in their atmosphere. The age and metallicity areas for the bulge and the halo are indicated. The three big pentagons with error bars are for globular clusters of the Sagittarius dwarf. From Haywood, M. (2009) in *Galaxies and Cosmology*, ed. Combes, F., Paris, Ellipses, p. 29.

We now turn to the bulge. As we saw in Chapter 5, the bulge, or rather the pseudo-bulge of our Galaxy, is currently powered by a migration of gas and stars due to the effect of the primary and secondary bars. Once the stars arrive in the bulge, they cannot get out. Thus the bulge accumulates all the stars that were brought there by the different processes encountered during the life of the Milky Way: accumulation due to several successive bars, or even stars captured from a companion galaxy. Some of these phenomena are relatively recent, but many of the stars of the bulge are older than 12 billion years. Among these old stars, some have completed their evolution on the main sequence and are currently at the stage of red giants; it is their metallicity that we can measure because they are the brightest. Given the relatively short duration of the episode during which they were formed, the approximation of instantaneous recycling in a closed box seems appropriate to describe their formation: actually, this approximation makes for a pretty good account of the distribution of metallicity in red giants of the bulge (Fig. 6.8). This distribution is then given by the equation:

$$dM_\star(Z)/d(\ln Z) = MZ/y_Z \exp(-Z/y_Z), \tag{6.4}$$

where $M$ is the total mass, $M_\star$ that of stars, $Z$ the metallicity and $y_Z$ the yield of heavy elements.

Note that the central region of the Galaxy contains metal-rich globular clusters, unlike those of the halo. Their stellar population is entirely similar to that of the bulge, indicating identical age, formation and evolution. Conversely, the metal-poor globular clusters in the halo are all old, and were formed from primordial gas only enriched by the Population III stars.

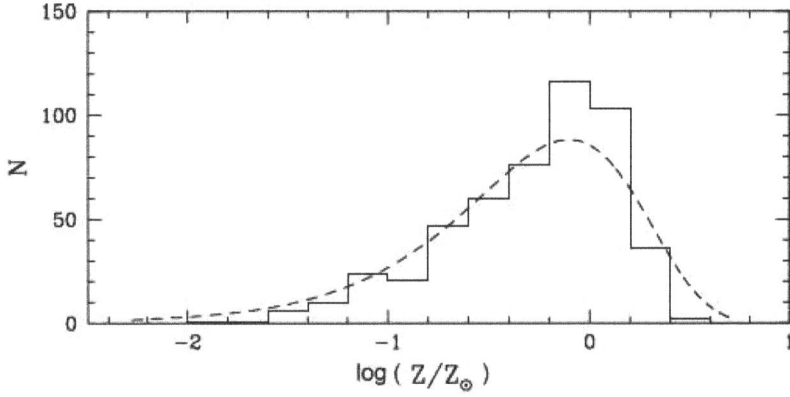

FIG. 6.8 – The distribution of metallicity for the red giants of the galactic bulge. The number $N$ of stars observed by logarithmic interval of metallicity is given as a function of the metallicity $Z$. The curve is the model prediction of the instantaneous recycling in a closed box, with a yield $y_Z = 0.015$. The agreement between the observations and the model is rather good. From Zoccali, M. et al. (2003) *Astronomy & Astrophysics* 399, 931-956, with permission of ESO.

*The chemical evolution of the Galaxy* 133

The [α/Fe] ratio is large for the old giant stars of the bulge (Fig. 6.9), which shows that the SNIa had not yet had time to explode in large numbers during the main period of stellar formation of this bulge, 12 to 13 billion years ago.

There is gas in the most central regions of the Galaxy, which is still forming stars. This gas contains deuterium, a hydrogen isotope that is destroyed in stars, which shows that it has not been completely ejected by stars of the bulge that destroy deuterium, but must come from elsewhere. The gas in the bulge partly comes from the disk, where deuterium has not been completely destroyed, as indeed was predicted by the dynamical models described in Chapter 5.

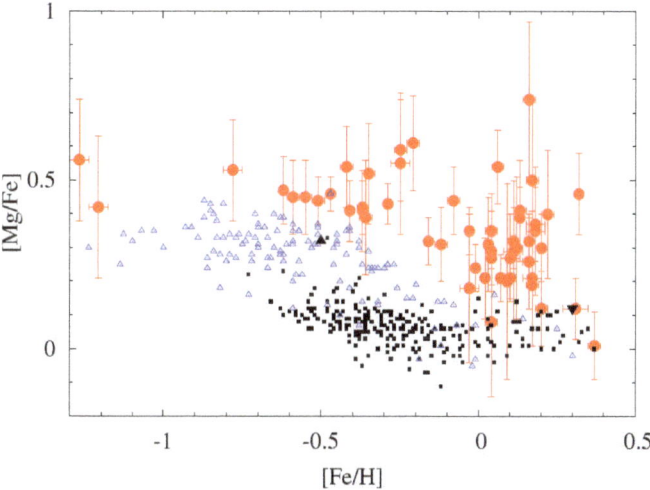

FIG. 6.9 – The abundance ratio [Mg/Fe] as a function of the abundance ratio [Fe/H] for stars of the galactic bulge (in red, with error bars), the thick disk (blue triangles) and the thin disc (black squares). Magnesium is an α element. The ratio is different for the three components, indicating different origins. The differences also indicate that the giant bulge stars are older than the dwarf stars of the thick disk, which themselves are older than those of the thin disk. The quantitative interpretation of this Figure is risky because the stars observed in the two disks are close to the Sun and are not representative of the whole disks. From Lecureur, A. et al. (2007) *Astronomy & Astrophysics* 465, 799-814, with permission of ESO.

## 6.5 The chemical evolution of disks

Unlike the halo and the bulge, the thin galactic disk was formed slowly and probably changed much since 8 to 10 billion years ago. The dating of stars, though difficult and relative, however suggests that the thin disc formed more stars in the last five billion years than in the previous period, although the overall rate of star formation in the whole Galaxy has probably decreased continuously from the maximum corresponding to the formation of the bulge. This result is surprising since the rate of star formation in the disc should be proportional to the mass of the gas available. This suggests that the gas was renewed by infall from the gas of the halo, the intergalactic medium and nearby galaxies: the closed-box model we have previously invoked for the bulge is not applicable to the disk. This is confirmed by the very small variation of the metallicity of stars in the disk in the last 8 billion years or so (see Fig. 6.7).

It is therefore natural to consider that low-metallicity gas coming from outside contributes to the evolution of the disk. In fact, we will see in Chapter 8 that the observations of the 21 cm line of hydrogen at high galactic latitudes show gas clouds, the so-called high-velocity clouds, which fall onto the galactic plane (see below Fig. 8.7). The external origin of this gas is confirmed by its metallicity, which is about one tenth of the solar metallicity. The mass of gas that falls at high velocity near the Sun covers about 10% of what is needed to maintain the current level of star formation. This is not much, but there is also gas falling at lower velocities, and it could be that the accretion of fresh gas actually compensates for the gas consumed by star formation, about 3 $M_\odot$ per year: if this is the case, half of the mass of the Galaxy (besides dark matter) originates from outside. This tends to be confirmed by the numerical models of formation and evolution of bars in galaxies (see Chapter 5). A large part of this gas can be extragalactic. Another part could come from very hot material ejected by the supernova and bubbles: This ionized gas at $10^6$ K would have cooled, recombined and fallen on the disk, possibly far from its place of origin: this is the *galactic fountain*.

The simplest models that we can build in this context assume that there is no radial transport of gas or stars in the Galaxy, the rings at different galactocentric radii evolving independently of each other. To account for the observations, these models assume that the thick disk formed first, during about 0.8 billion years, in part from external primordial gas. Then the creation of stars would be interrupted as a result of some violent event, for example the infall of a satellite galaxy. The thin disk would then have formed slowly from gas enriched in heavy elements coming from the thick disk but also from gas from the intergalactic medium and from the remaining gaseous halo. The internal parts of the thin disk would have appeared first, then this disk would extend to the outer regions. This formation would have lasted for 7 billion years at the solar radius. These models reproduce the characteristics of the disks, the distribution of metallicity of stars near the Sun, and the abundance gradients rather well.

However, they do not take into account the radial migration of stars and gas in the disks (see Fig. 5.5). The gas tends to move toward the inside of the galactic disk, at least in the regions interior to the co-rotation. As for the stars, they are mixed radially in both directions while their velocity dispersion increases slightly over time: the interactions "heat" all stars, so that the disk of gas and young stars is thinner than the disk formed of the oldest stars. 1 to 3 billion years after their formation, the trajectories of stars lose any information about their initial state. The thin disk thus results from a mixture of stars of different mass, age and chemical composition. This explains the absence of correlation between metallicity and age of stars in the solar neighborhood (see Fig. 6.7). To account for this, the models without radial transport must assume that a continuous infall of gas on the disk dilutes the newly formed heavy elements by the appropriate amount of low-metallicity gas, which is somewhat artificial. It seems that one is led to abandon these models, especially as the radial transport is an inevitable consequence of the presence of a bar and spiral arms.

Figure 6.10 shows the relationship between the rotational velocity and the metallicity [Fe/H] or the abundance ratio [$\alpha$/Fe] for different stellar components of the Galaxy. All the stars marked with dots are near the Sun. It is found that the thin disk stars have rotational velocities around the galactic center between 185 and 230 km/s (here the rotation velocity of the LSR is taken as 220 km/s). They therefore have orbits that may appreciably differ from a circular orbit, albeit moderately. However, the stars of the thick disk have smaller rotation velocities at the solar radius, and therefore far more eccentric orbits. Their average metallicity is lower than that of the thin disk, with a larger dispersion. Moreover, the [$\alpha$/Fe] ratio is larger and more dispersed for the stars of the thick disk than for the thin disk, and many have a [$\alpha$/Fe] ratio similar to that of halo stars. All this shows that the stars of the thick disk are on average much older than those of the thin disk, and have been more subject to gravitational perturbations. We also suspect, considering Figure 6.10, that the mean metallicity of the stars of the thin disk decreases somewhat when the velocity increases, while the opposite is true for the stars of the thick disk. These suspicions are confirmed by recent studies.

Note that the models that attempt to reconstruct the history of star formation in the solar neighborhood from observation do not show the exponential decay with time of the formation rate predicted by closed-box models (Fig. 6.11). An accretion of fresh gas throughout the life of the Galaxy is thus supported by these observations. Some of these models suggest the occurrence of violent events, which resulted in an increase of star formation.

Recent models taking into account the radial migration of gas and stars succeed to account for all these phenomena in a simple way, without any particular assumption on a stopping of the star formation in the thick disk after some violent event followed by a new formation in the thin disk, a necessary hypothesis in models without migration: the thick disk and thin disk appear naturally in models with migration without any such discontinuity

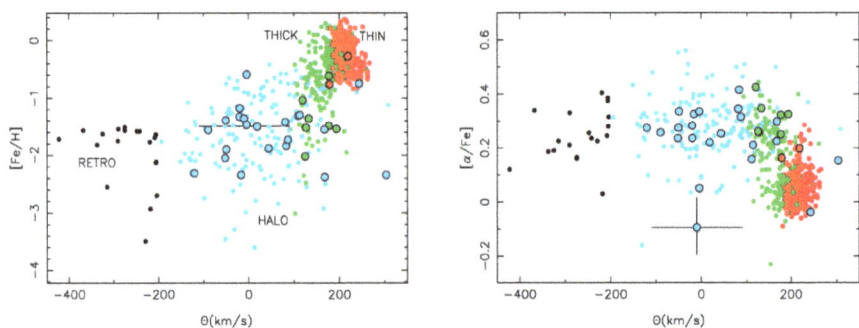

FIG. 6.10 – Variations with the rotation velocity of the metallicity, namely [Fe/H], and of the abundance ratio [α/Fe] for different stellar components of the Galaxy. The circles correspond to globular clusters (filled circles, blue for the halo clusters, and red for clusters whose kinematics is similar to that of the disk); the blue and black dots correspond to individual stars of the halo observed near the Sun (black for extreme backwards rotating stars), the green points correspond to the stars of the thick disk and the red ones to the thin disk stars, all in the vicinity of the Sun. From Pritzl, B.J., Venn, K.A. & Irwin, M. (2005) *Astronomical Journal* 130, 2140-2165, with permission of the American Astronomical Society.

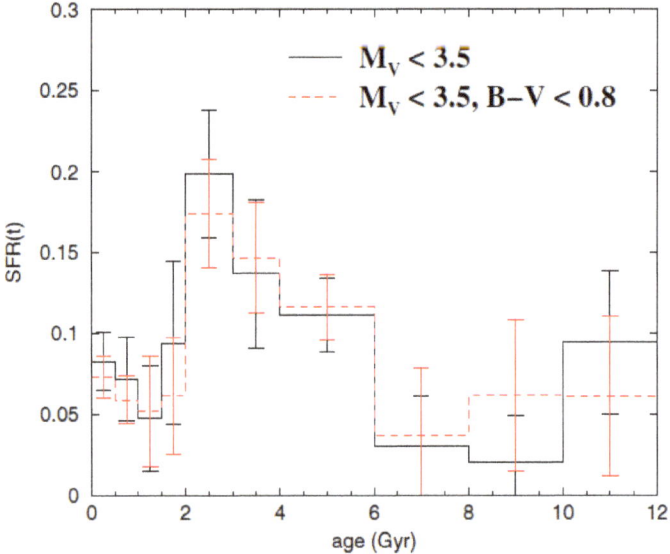

FIG. 6.11 – Reconstruction of the history of star formation near the Sun, derived from a complete sample of stars brighter than the absolute magnitude $M_V = 3.5$ (solid line), and with color B-V < 0.8 (dotted line). From Cignoni, M., et al. (2006) *Astronomy & Astrophysics* 459, 783-796, with permission of ESO.

in the star formation rate. These models will certainly develop further and evolve as they still contain many arbitrary parameters, such as the infall rate of gas and its distribution over the disk: for the moment, we can only determine these parameters by adjusting the results of the calculations to observations. Many observations are still needed, because the stellar population is only well known in the vicinity of the Sun, and is the result of a mixture of more distant populations. There is still much to do to really understand the evolution of the Galaxy, whose complexity seems to increase as we know it better. Again, the GAIA satellite should allow great progress in this field.

# Chapter 7
# Formation and evolution of the Galaxy

Our Galaxy can be divided into three luminous components, which have very different properties and were formed at different times by different mechanisms: the disk, which contains more mass in both stars and gas, the stellar bulge and the halo, much less massive with older stars on average. The disk itself is divided into two components, the thin disk and the thick disk; the thin disk is the youngest and the most gas rich.

## 7.1 The thin and thick disks

As we saw in Chapters 3 and 6, the thin disk contains the gas and the young stars, rich in metals, and the thick disk consists of older stellar populations whose metallicity is lower, without reaching the extremely low values of the stellar halo. It therefore appears that the thick disk was formed before the thin disk, and that new stars are now only formed in the thin disk.

Several scenarios could account for these differences. The thick disk could have been formed by:

1) accretion of stars belonging to small satellite galaxies destroyed by tidal effects, accretion which also accounts for tidal streams that exist in large numbers in the stellar halo;

2) heating of the thin disk by interactions and minor mergers with companions, this time more massive;

3) radial migration of stars by scattering at the co-rotation resonance with spiral waves or bars, as described in Chapter 5;

4) formation *in situ* at the beginning of the Universe, from a disk of gas thickened by gravitational instabilities or interactions.

It is possible that all these mechanisms have played a role in certain epochs, and today we observe the accumulation in the thick disk of all the resulting stars. But how can we know which processes dominated?

For the first mechanism, numerical simulations confirm that stars from small satellites can indeed spread over the galactic plane, since they are progressively stripped with lower energy due to dynamical friction. However many satellites may have more polar orbits, so that tidal streams are found in the halo, or even end up in retrograde rotation from the original sense of rotation of the disk stars. This does not match with what is observed: the stars of the thick disk rotate in the same sense as in the thin disk, even if they do not rotate as rapidly. Moreover, the orbits of the stars formed in this scenario would be much more eccentric than observed. In general, the eccentricity of the orbits is one element that distinguishes between the various scenarios, as shown in Figure 7.1.

In the second mechanism, the thick disk comes from the thin disk, at least from the stars it had a few billion years ago, when a massive companion would have interacted with the Milky Way, and heated and disturbed the thin disk. The stars of the disk at that time were less rich in heavy elements, which could explain the current characteristics of the thick disk. The companion should have been at least 20% as massive as the Galaxy, and would eventually merge into the bulge. The main problem with this scenario is that the bulge of the Milky Way is not so massive, and also the stellar orbits of the thick disk should be more eccentric than what is observed.

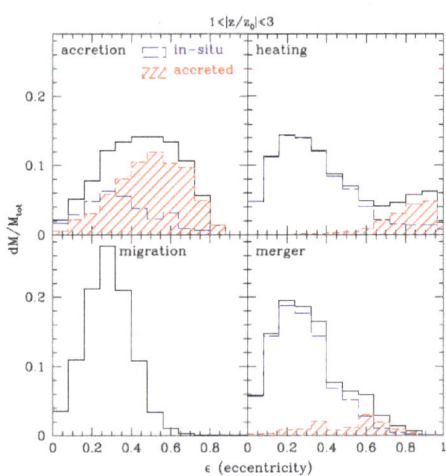

FIG. 7.1 – Comparison between several models of the formation of the thick disk. For each model, the distribution of the eccentricities of the stars is indicated, when their height above the plane is between 1 and 3 scale-height $z_0$, and the distance from center between 2 and 3 radial scales $R_d$. The red-shaded areas correspond to accreted stars, areas under the thin line to those formed *in situ*; the total is in bold lines. The models considered are from left to right and top to bottom, the accretion of small satellites, the heating due to galaxy interactions, the radial migration, and mergers between galaxies. From Sales, L.V. et al. (2009) *Monthly Notices of the Royal Astronomical Society* 400, L61-L65, Wiley, with permission of the publisher.

The third scenario also forms the thick disk from the thin disk, but with stars that migrated at different times, mainly from the center of the Milky Way. This mechanism must in any case exist and add to all the other scenarios because the Galaxy has spiral arms and a bar that cause these migrations, and certainly had several density waves during its lifetime, with several pattern speeds. The average metallicity of the thick disk stars that come from this migration is then less than the current thin disk metallicity, and the rotation of the thick disk is significant and in the same sense as that of the thin disk, as is actually observed (Fig. 7.2). It must be assumed in this scenario that after such events the galaxy accretes again external gas to reform the thin disk.

Finally, in the fourth scenario, the stars of the thick disk were formed *in situ* from a disk of gas that was very thick in the past. This thick disk could come from gas accretion during a merging between galaxies; more simply, the Galaxy early in its life was very rich in gas, thus very unstable, which would have made the gas very turbulent, forming large gaseous fragments with an important velocity dispersion, which would have thickened the disk. These events occurred early in the evolution of the Galaxy when the metallicity was still very low. Some of these fragments must have also, by dynamical friction, migrated toward the center and contributed to the formation of the bulge.

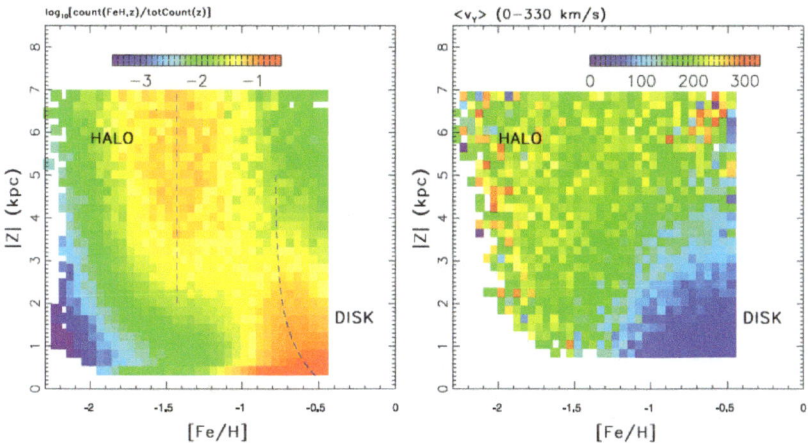

FIG. 7.2 – Left: distribution of stars of the Milky Way, metallicity ([Fe/H]) vs height above the plane. The dotted lines show that the halo stars have a low metallicity, but are constant with height, while the disk metallicity decreases from the thin to the thick component. On the right, in the same diagram, the color this time represents the rotational velocity *relative to the Sun*: A value of 220 km/s means that the stars do not rotate. We see that the stars of the thick disk rotate in the same sense as the thin disk, but with strong differences from one star to another, while the halo does not rotate. From Ivezić, Ž. et al. (2008) *Astrophysical Journal* 684, 287-325, with permission of the American Astronomical Society.

## 7.2 The formation of the bulge

In many spiral galaxies, the bulge is a stellar component with an ellipsoid shape, which resembles an elliptical galaxy in miniature. Such a "classical" bulge does not rotate, is not very flattened, and has a light distribution varying with the radius $r$ as $\exp(r^{1/4})$ (the de Vaucouleurs's law). Its color is red. The bulge of the Milky Way is quite different: it is a pseudo-bulge. It is rather flattened, has a peanut shape, an exponential radial distribution (such as disks), a blue color and it is rotating. These features suggest that many of the bulge stars come from the disk by secular evolution, as we have described in Chapter 5.

The first way to form bulges, especially the classical bulges, is by the merging of galaxies. Just like major mergers between spiral galaxies form elliptical galaxies, a merger between a spiral galaxy and a smaller mass companion will result in stellar accumulation in a bulge. There may be several successive mergings, and since the angular momentum of the different merging satellites has a random direction, the rotation eventually disappears. It is possible that a fraction of the bulge of the Milky Way has formed like this, but this process cannot be dominant.

The second scenario is the stellar migration of disk stars in the bulge due to resonance with the bar and subsequent secular evolution. This process was described in Chapter 5, and results mainly in box or peanut-shaped bulges according to the orientation along the line of sight. The stars retain much of their angular momentum of rotation, and younger stars can still be driven into the bulge, resulting in a bluer color than that of classical bulges. The radial distribution of mass retains the exponential form it has in the disk. The bulges formed in this way are not as massive as classical bulbs, and this corresponds to the case of the Milky Way pseudo-bulge.

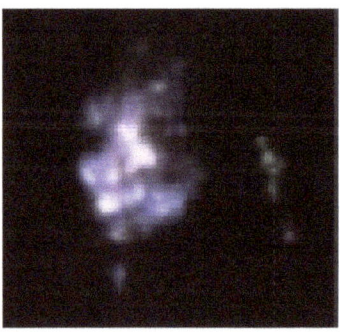 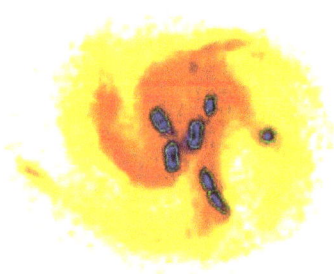

FIG. 7.3 – Left: example of clumpy galaxies observed to be more numerous at high redshift, soon after the beginning of the Universe. On the right, simulation of the formation of these clumpy galaxies from a disk whose mass is 50% gas: the gravitational instabilities are so strong that the gas fragments in giant clumps. From Bournaud, F. et al. (2007) *Astrophysical Journal* 670, 237-248, with permission of the American Astronomical Society.

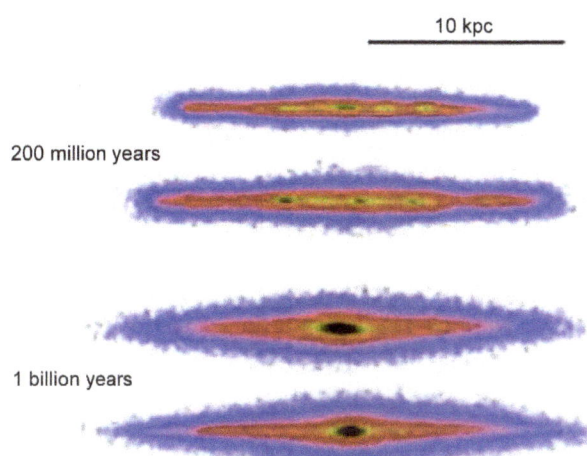

FIG. 7.4 – In the same simulation as for the previous figure, we see now the galaxy edge-on and its evolution over time, in two perpendicular directions. The clumps in the plane now appear aligned. Gradually, by dynamical friction, they spiral towards the center to form a bulge, and the process lasts just a billion years.

Finally, there is a third way of forming bulges, very early in the Universe, from the giant gaseous clumps formed in some galaxies at that time, where more than half of their mass was made up of gas. These clumps result from instabilities of the gaseous disk, instabilities that will also form stars efficiently before they can even generate spiral waves. We have seen that this is a possible scenario for the formation of the thick disk. These clumps are so massive that the dynamical friction against the rest of the disk and also the dark matter will slow them down and quickly drive them to the center (Figs. 7.3 and 7.4). The stars that populate these clumps will then end in a classical bulge of low metallicity and high ratio [$\alpha$/Fe] as the stars formed quickly. The scenario is possible, but it involves a lot of uncertainty because violent star formation in the clumps may destroy them through stellar winds and supernova explosions before the clumps arrive in the center.

The three scenarios to form bulges that we have just described are not always effective, because the number of galaxies devoid of a bulge, or with a very small bulge which is usually a pseudo-bulge, is very significant and even dominant. A large number of spiral galaxies have not been subject to major mergings over billions of years, otherwise they would all have a massive classical bulge. This is probably the case of the Milky Way. The fraction of galaxies where the result of disturbances and mergers can be observed was recently estimated as about 10% during the second half of the Universe life. So it seems that most galaxies acquire their mass by the accretion of intergalactic matter rather than interaction and merging with other galaxies.

## 7.3 The formation of the halo : cosmological or not?

The first ideas about the origin of the stellar halo of the Milky Way assumed a global progressive collapse of an initial mass of gas. In the 1960s, Olin Eggen (1919-1998), Donald Lynden-Bell and Alan Sandage (1926-2010) had developed this scenario, since then called ELS, based on the observation of the kinematics and the heavy elements abundance in the halo stars. The halo stars are animated by disordered motions, are not on average rotating around the galactic center, their orbits are eccentric, and their heavy elements abundance is low. These observations could be reproduced by assuming a gas cloud with almost primordial abundance, which would form stars gradually while collapsing, flattening and concentrating, thus rotating faster and faster around the center. We then expect to find a vertical abundance gradient, as well as ages getting younger when approaching the disk plane: today young stars form in the disk where all the gas is confined.

However, in the 1970s, critics did notice that the halo had many globular clusters, with various abundances independent of their distance to the center, and favored a stellar halo formed by interactions with companions, leading to accretion of external matter. For ten years, many different stellar streams were observed in the halo, which support the hypothesis of accretion of small companions, and galaxies in the process of being destroyed by tidal interaction exist, such as the Sagittarius dwarf galaxy (see Chapter 6). The question now is whether this mechanism dominates all the others, or if there are also some in the halo stars that come from a global collapse as in the ELS scenario.

One diagnostic may be the degree of inhomogeneity and of anisotropies in the stellar halo, which characterizes the tidal streams. This degree was estimated at about 50%, which means that at least half of the stars have a homogeneous distribution. However, they may not all come from a global collapse because the streams can relax and homogenize, quicker near the center of the Galaxy: simulations are necessary to quantify the remaining irregularities.

Such simulations have been performed by several groups, as shown in Figures 7.5 and 7.6. In these simulations, the Milky Way is formed in a cosmological context, by accretion of cosmic filaments and merging of smaller galaxies. Simulations show much more substructures than observed, which probably comes from the large number of satellites that appear in the models, although most of them are supposed to have no stars. Anyway, these simulations suggest that only about 10% of halo stars were formed *in situ*, and do not come from external accretions.

In conclusion, the recent observations of large stellar surveys have brought a wealth of information about the possible origin of the Milky Way. The discovery of many tidal streams reflects the accretion of multiple satellites, which formed the bulk of the stellar halo, and perhaps some of the thick disk. The detailed observation of the kinematics of the stars, in addition to the observation of heavy element abundances, including the ratio [α/Fe], allows us to date the various populations, and to speculate about their history and dynamics. In particular, the eccentricity of the orbits of the stars could be a decisive criterion for choosing among the various possible scenarios. However, the secular evolution blurs the picture, mixing stars radially and vertically: those born to the center can find themselves on the outside, and this is probably the case of the Sun, which was not born at its current galactocentric radius. Numerical simulations still help to distinguish between the various possibilities. Soon the GAIA satellite will measure the distance, the proper motion and radial velocity of a very large number of stars of all distances in the Milky Way, for which we will trace the three-dimensional dynamics with much more certainty.

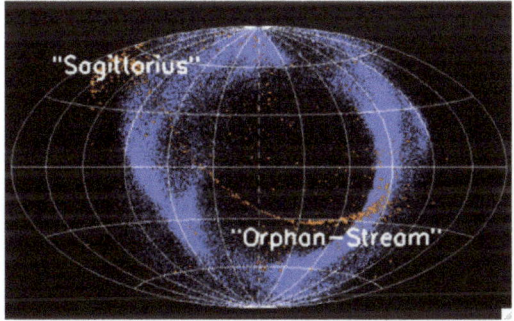

FIG. 7.5 – Distribution in the sky of two tidal streams, similar to those in the Sagittarius dwarf and Orphan stream, but from a numerical simulation of the Milky Way interacting with its satellites. From Helmi, A. et al. (2011) *Astrophysical Journal* 733, L7, with permission of the American Astronomical Society.

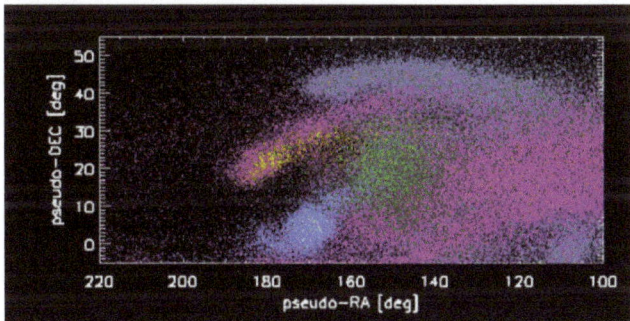

FIG. 7.6 – In the same simulation as the previous figure, representation of tidal streams in equatorial coordinates on the sky. The stars of the various streams are shown with a different color, and we see that several satellites can contribute to the same apparent stream. The interpretation of the observations is therefore more complex than expected.

# Chapter 8
# The Galaxy among its companions

Our Galaxy is part of the Local Group, of which the most massive elements are the Andromeda Galaxy (M 31) and the Milky Way. In this, our Galaxy is no exception, because the majority of galaxies live in groups. Closer to our Galaxy, are the Magellanic Clouds, and a large number of small satellites of which many, very faint, have been discovered only recently. Finally, our Galaxy is not a closed system but is connected to its environment by gas accretion from cosmic filaments. It is possible that the high velocity clouds of hydrogen gas, which mostly fall on the plane of the Galaxy, are a manifestation of this continuing accretion.

## 8.1 A spiral among the spirals – the Hubble classification of the Galaxy

How is our Galaxy situated among all types of galaxies observed in the Universe? The Milky Way is a relatively massive spiral galaxy. The luminosity function $F(L)$ (and consequently the mass function) of galaxies follows the so-called Schechter relation, which is a power law up to a typical luminosity $L^*$, then decreases exponentially. $L^*$ is the value of the turn-over of the distribution (see Fig. 8.1).

The Schechter function is written as $F(L) = F^*(L/L^*)^a \exp(-L/L^*)$, where $F(L)$ is the density of galaxies per unit volume (Mpc$^3$) and unit luminosity. The values of the slope are on the order of $-1$, and the critical luminosity $1.5\text{-}2 \times 10^{10}$ $L_\odot$. Although small galaxies are more numerous, the luminosity in the Universe is dominated by galaxies with luminosity close to $L^*$.

The Milky Way also obeys the Tully-Fisher relation, which correlates the total luminosity (or the total mass of baryons) to the rotational velocity, as shown in Figure 8.1. We have already seen this relationship in Chapter 3, which allows us to better understand the relationship between dark matter and visible matter in galaxies. Again, the position of our Galaxy shows that it is a giant galaxy, one of the two most massive of the Local Group.

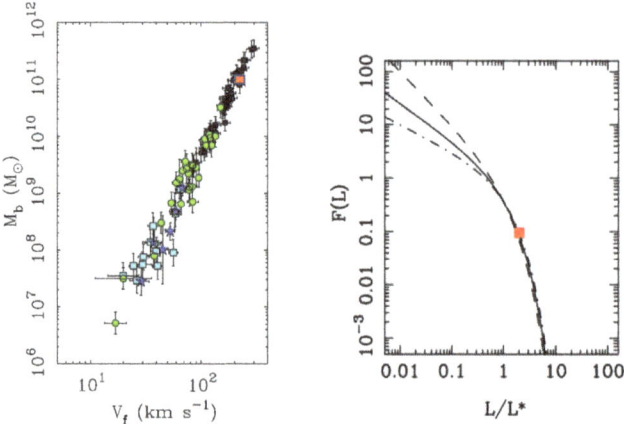

FIG. 8.1 – Left, the baryonic Tully-Fisher diagram for spiral galaxies. The total mass of baryons $M_b$ (stars plus gas) is plotted as a function of the rotational velocity $V_f$ when it reaches a plateau. The position of the Milky Way is indicated by the red square, with $V_f = 220$ km/s, $M_b = 10^{11}\,M_\odot$ (adapted from McGaugh S.S., 2012, *Astronomical Journal* 143, 40, with permission of the American Astronomical Society). See also Figure 3.33, where the Milky Way is located at the upper right of the diagram. On the right, the Schechter luminosity function, with $a = -1$ (dashed line), $-0.7$ (full line), and $-0.5$ (dot-dash). The red square represents the Milky Way.

In terms of its morphology, our Galaxy is a spiral of rather "late" type (in the sense of an old nomenclature that does not have much interest today), which still has a developed disk of gas and young stars (Fig. 8.2). Its bulge is not very significant in relation to its disk, unlike its neighbor the Andromeda galaxy.

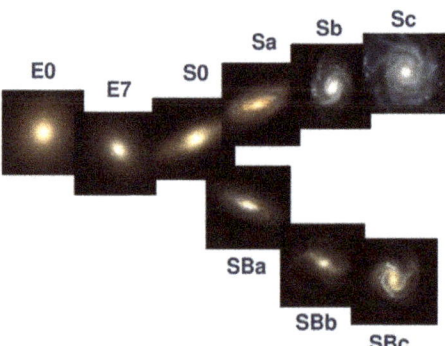

FIG. 8.2 – The Hubble classification, built with pictures of galaxies of the various types. The two branches of the "tuning fork" are the "normal" spiral galaxies S and barred SB, respectively. The Milky Way is a barred galaxy of intermediate type between SBb and SBc.

The bulge of our Galaxy, seen in near infrared, has a box or peanut shape, which corresponds to a bulge coming mainly from internal secular evolution, from a bar. Around the Andromeda galaxy, which has a much more massive and extended bulge, one has observed stellar debris and remains from a major merger. It seems that our Galaxy in comparison had a quieter past and has not undergone any major merger over the past 6 billion years.

## 8.2 The satellites : the Magellanic Clouds and dwarf elliptical galaxies

The Magellanic Clouds, the Large one (LMC for "Large Magellanic Cloud") and the Small one (SMC, "Small Magellanic Cloud"), are two dwarf galaxies that orbit the Milky Way, at a respective distance of 50 and 63 kpc. Observable only from the southern hemisphere, they were first identified in visible light, from their stellar component. Like any galaxy of "irregular" type, they are rich in atomic gas, and their emission in the 21-cm line unveiled their very advanced interaction with the Milky Way. This interaction results in highly developed tidal tails, which surround the galactic plane as a polar ring called the Magellanic stream (Fig. 8.3).

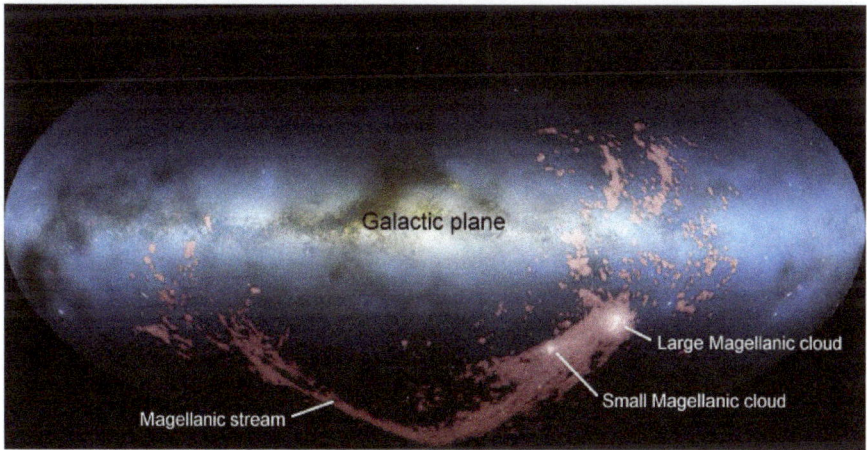

FIG. 8.3 – Overlay of the HI 21cm emission from the Magellanic Clouds, and Magellanic stream (pink) to an optical image of the Milky Way, seen in the galactic latitude and longitude reference frame, with the center of the Galaxy at the image center (from Nidever D.L. et al. 2010, *Astrophysical Journal* 723, 1618-1631, with permission of the American Astronomical Society).

The evidences of the interaction of the Magellanic Clouds with the Milky Way are multiple, and many models have been developed over the years, with significant variations. Early models assumed that the Magellanic Clouds were bound for more than an orbit to our Galaxy, and that they were at least in their second passage, if not more. The radial velocity of the Clouds observed by the Doppler effect is indeed only of the order of 100 km/s, much less than the escape velocity, and the system should then be gravitationally bound. Yet, more recently, the velocity of the Clouds in the plane of the sky could be estimated through the proper motions of stars they contain, estimated from successive images taken by the Space Telescope. The total velocity was then found equal to 378 km/s, slightly higher than the escape velocity from the Milky Way. The Clouds would then be only now entering for the first time into the galactic system, and thus would just be about to be captured by the Milky Way.

The uncertainties in this problem are still large, and the two hypotheses remain today. The dynamical friction, which slows down the satellites as they enter the system, and changes orbital energy to internal energy, can be very effective; but it depends on many unknown parameters of the Clouds orbit, and also their degree of mass concentration before their collision with the Milky Way. While it is easy to explain the development of the Magellanic stream by tidal tails that typically grow during a hundred millions years, with several passages of the Clouds, it is more difficult to account for it if the Clouds are not yet bound to the Milky Way. The only possible hypothesis would be the action of the ram pressure of the hot intergalactic medium, which strips the neutral gas of the galaxies when they pass through this medium at high velocity. The ram pressure is proportional to the density of the hot intergalactic gas and the square of the relative velocity. The problem with this hypothesis is that the density of the hot gas is very low, and that the action time-scale of the ram pressure cannot have been very long if the Clouds pass through this hot gas for the first time. In addition, the Magellanic stream is not composed of only one trail behind the Clouds, but also of a component leading forward on their trajectory, which is impossible to be explained by the ram pressure. We must then assume that the long trail of atomic gas was formed well before the entry of Clouds in the sphere of influence of the Milky Way, through the mutual tidal interaction between the two dwarf galaxies SMC and LMC. This interaction was simulated, and helps to explain some of the deformations of these dwarf galaxies, including the fact that the LMC has a significantly eccentric stellar bar.

In addition to the Magellanic Clouds, a large number of small dwarf galaxies, spread in all directions around the Milky Way, appear to be gravitationally bound to our Galaxy: among others, the Sagittarius dwarf (Sag DEG), the Canis Major dwarf, which is deeply disturbed and stretched by tidal forces and could even be just a concentration of galactic stars, and the more distant dwarf galaxies Ursa Minor and Major, Draco, Carina, Sextans, Sculptor, Fornax Leo I, Leo II, Phoenix and perhaps Leo A (which belongs to the Local Group, but could be part of another subgroup than the Milky Way), as well as Antlia, Cetus and Tucana, which seem more isolated. More recently a dozen very faint dwarf spheroidals have been discovered, thanks to the large sky surveys at high sensitivity. The search done with the SDSS (Sloan Digital Sky Survey) helped us to significantly advance the topic. In the Appendix, one can find the current list of the brightest satellites of the Milky Way, with their characteristics. Our neighbor the Andromeda galaxy has even more satellites.

Although these satellites are numerous, those that are currently being discovered are faint, and are even ultra-faint with very little mass. And for a given mass and luminosity, the number of satellites is not nearly as large as what is predicted by the standard cosmological model, which is based on the existence of cold dark matter (CDM). Cosmological simulations indeed predict that the number of satellites around a galaxy of the mass of the Milky Way should be several thousands! Figure 8.4 shows the result of such a simulation of a dark matter halo that could match that of the Milky Way, surrounded by its satellites. The diagram shows the abundance of satellites as a function of mass, and why the ultra-faint satellites that can still be discovered today or tomorrow will not solve this problem. This is called the missing satellites problem. A solution would be that all these small satellites are constituted exclusively of dark matter, without stars. Yet all the faint and ultra-faint satellites recently discovered have all too many stars to solve the problem.

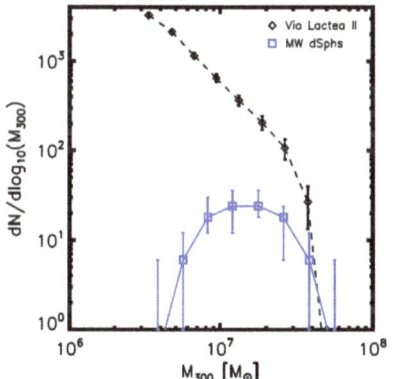 

FIG. 8.4 – Left: diagram of the number of satellites predicted by the CDM cosmology model (black dotted line) as a function of the mass $M_{300}$ of dark matter included within a radius of 300 pc. This prediction can be compared to the observations, the full blue line which also makes the census of the satellites with the same mass (from Bullock J.S. 2010, arXiv: 1009.4505). Right: simulation of a halo corresponding to the Milky Way (from the simulation Via Lactea II of Diemand J. et al. 2008, *Nature* 454, 735-738, thanks to J. Diemand).

## 8.3 Capture of the Sagittarius dwarf, and many others: the tidal streams

The Sagittarius dwarf is the closest companion to the Sun, at 24 kpc; this galaxy is in the process of being destroyed and stretched by tidal interaction with the Milky Way. We observe two stellar arms upstream and downstream of the orbit, which are spread out filaments covering more than 100 degrees on the sky (Fig. 8.5). The Sagittarius dwarf is located on the other side of the galactic center from the Sun, and therefore is difficult to observe: it was discovered in 1994, and some of its stars, especially carbon stars, were considered for a while as stars of the galactic bulge. In the filaments, there is also gas that participates in the tidal stream. The tidal debris could contain more than 70% of the mass of the original galaxy, but the models are uncertain and depend on when the galaxy collided with the Milky Way. One thing is certain, though: in about 100 million years, the Sagittarius dwarf will fall back on the plane of our Galaxy, and end up being absorbed.

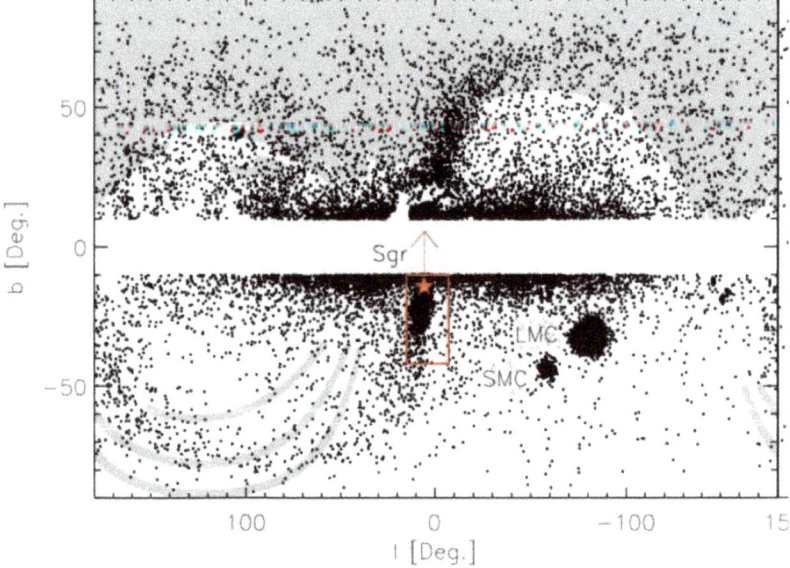

FIG. 8.5 – Representation in galactic coordinates (the plane of the Milky Way is masked in white around $b = 0$) of all M giant stars belonging to the Sagittarius dwarf. One clearly distinguishes the upstream tidal arm (top) and downstream (bottom) of the trajectory. The center of the dwarf is indicated by the red star, it moves upward along the red arrow. Also note the presence of Magellanic Clouds (LMC and SMC). From Niederste-Ostholt et al. (2010) *Astrophysical Journal* 712, 516-526, with permission of the American Astronomical Society.

The Galaxy among its companions 153

Although many data have been accumulated on the Sagittarius dwarf since 1994, (notably proper motions thanks to the Space Telescope, chemical abundances, star clusters and even globular clusters belonging to the stream), its past trajectory is still uncertain and estimates of its initial mass vary by one to two orders of magnitude. In the center of the dwarf is the globular cluster Messier 54, and it seems that a high density of stars and a black hole of mass 10 000 $M_\odot$ exist there.

Accurate simulations were made of possible past trajectories for the Sagittarius dwarf (Fig. 8.6). They could help to determine the three-dimensional shape of the dark matter halo of our Galaxy, if the stars in the stream were in dynamical equilibrium, which is not certain.

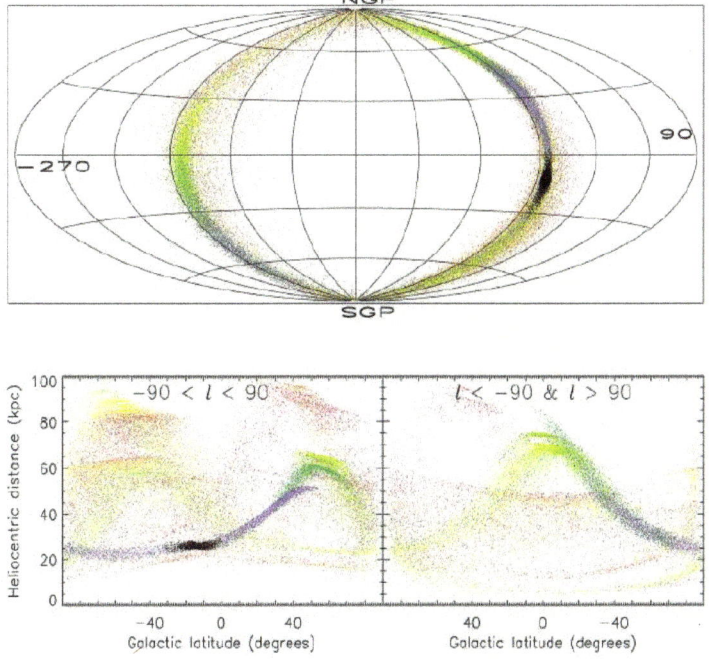

FIG. 8.6 – Simulation of the interaction between the Sagittarius dwarf and the Milky Way, and formation of the tidal stream. The stream is spread in a polar ring, viewed in galactic coordinates, at the top. The stream is then represented with the same colors in other coordinate systems. The nucleus of the Sagittarius dwarf galaxy is in black. From Helmi, A. & White, S.D.M (2001) *Monthly Notices of the Royal Astronomical Society* 323, 529-536, Wiley, with permission of the publisher.

As we discussed in Chapter 3, many other tidal streams were discovered in the halo of the Milky Way, and it could be that past interactions with dwarf galaxies were at the origin of all halo stars. A tidal stream was even identified as a ring in the outer parts of the galactic plane, the Monoceros ring (see Fig. 3.7), which could have come from the Canis Major dwarf. The ring is wound 3 times around the Galaxy, and merges with the warped part of the plane.

From 2017, the GAIA astrometry satellite should give us proper motions for the brightest stars of the two Magellanic Clouds and the closest dwarf galaxies, like that of Sagittarius. This should allow us to significantly improve our understanding of our Galaxy interaction with these systems.

## 8.4 Galactic wind, high velocity clouds, cosmic accretion

Our galaxy is not isolated, but benefits also, as was the case throughout its life, from intergalactic gas accretion; this maintains the rate of star formation almost constant, rejuvenates the disk, and re-generates spiral or bar instabilities. The accretion rate is still not well known but could be of the order of a solar mass per year, or a little more. The gas falls in a too diffuse form to be detected at large distances from the galactic plane, but is spread over huge volumes, thus corresponds to a significant mass. When the gas condenses into clouds, it can be detected in the 21-cm HI line: this is the phenomenon of high velocity clouds, which has been known for more than half a century. These clouds (Fig. 8.7) all have a negative radial velocity, thus they are approaching the Galaxy, with the exception of those in the Magellanic Stream, which we have already mentioned.

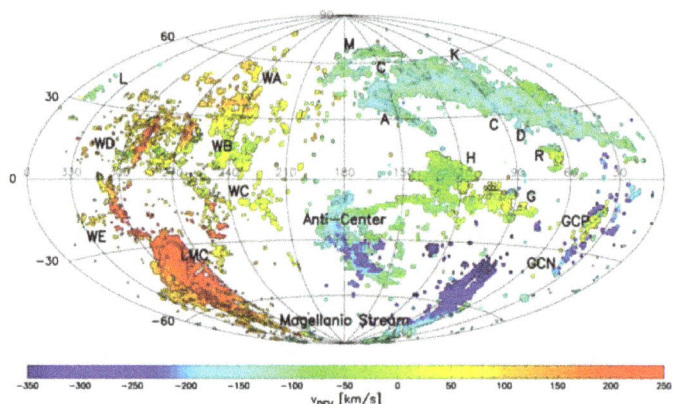

FIG. 8.7 – Map of the atomic hydrogen falling at high velocity on the galactic disk near the Sun. The coordinates are Galactic, the anticenter being in the middle of the figure. The color (bottom scale) indicates the excess (positive or negative) of the radial velocity with respect to extreme ones permitted by the differential rotation in the direction considered. We recognize at the bottom the Magellanic Stream. The other complexes are designated by letters. From Sancisi, R. et al. (2008) *Astronomy & Astrophysics Reviews* 15, 189-223, Springer, with permission of the publisher.

Do we see a similar phenomenon in nearby spiral galaxies? Many observations have been trying to detect clouds at high velocity with the HI 21-cm line: this is a difficult job, given the low brightness of these clouds. In spite of this, such clouds have been detected; they may be far away from the disk of their galaxy, for example in NGC 891, which is a galaxy very similar to the Milky Way, viewed edge-on (Fig. 8.8).

What is the kinematics of these clouds? If they do not all fall at high velocity, they have a different kinematics to what happens in the plane. They show a rotation around the galactic center, but with a lower speed than the rotation velocity in the plane. The further away from the plane they are, the less important the rotation is. These general features have helped us to build models of the origin of the gas. It is possible that part of the gas comes from the "fountain effect": it had been expelled by the young stars in the plane, either through supernova explosions or through stellar winds. Part or all of this gas does not reach the escape velocity out of the Galaxy, and, after cooling, rains back on the disk: this is what is referred to as the Galactic fountain. Yet not all the gas can come from this, because the energy required to raise all this gas at this height would be several orders of magnitude larger than the energy expanded in the formation of stars and supernovae. A mixture of scenarios must therefore exist, with an external gas accretion rate of about a few solar masses per year. This pattern also corresponds well to the observed gradient of rotation velocity in the vertical direction. If the gas above the plane came only from the gas ejected from the disk, its rotation should be much stronger. Several models have been developed and compared with observations. The picture that emerges is similar to the mechanism described in Figure 8.9, wherein both the gas ejected from the disk and the gas accreted from intergalactic space contribute to the observed gas in the halo. The ram pressure of the ejected gas could even help the gas arriving from outside to condense and form observable clouds. Then there would be a sort of cycle with motions in both directions, in an interaction zone.

The presence of gas in the halo is much more complex than it appears at first glance. It is certain that we are always in the presence of several gas phases, covering a wide range of densities, temperatures and pressures, and whose equilibrium is never reached. The fact that all gaseous disks of spiral galaxies, and not just that of the Milky Way, are deformed and warped (see Chapter 3) is just a consequence of the continuous accretion of intergalactic gas. The gas arrives at any random orientation, generally inclined relative to the axis of rotation of the disk galaxy. Once captured, it will rotate in an inclined plane in the outer parts, and give to the disk the appearance of a warped and moving pancake. Gradually the newly arrived gas will align through differential precession in the equatorial plane, which will also tilt slightly, as the angular momentum of the ensemble is constant. Progressively the plane of the galaxy is rotating, and one estimates that a complete turn, where the galaxy loses its initial orientation, occurs in 7 to 10 billion years.

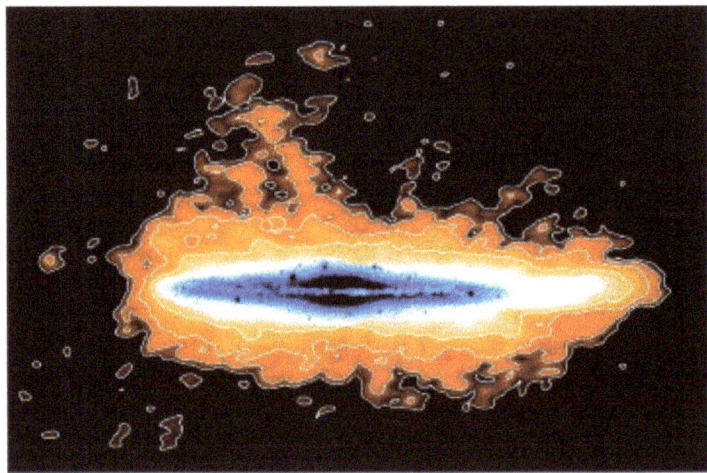

FIG. 8.8 – Contours of the HI emission of atomic gas at 21cm in red, overlaid on an optical image of the edge-on galaxy NGC 891. The gas spreads up to 25 kpc above the plane of the galaxy. The halo contains about 30% of the total gas content of the galaxy. These deep observations have required 200 hours of observations with the Westerbork radio telescope, in the Netherlands. From Fraternali, F. et al. (2004), *ASP Conference proceedings* 331, with permission of the American Astronomical Society.

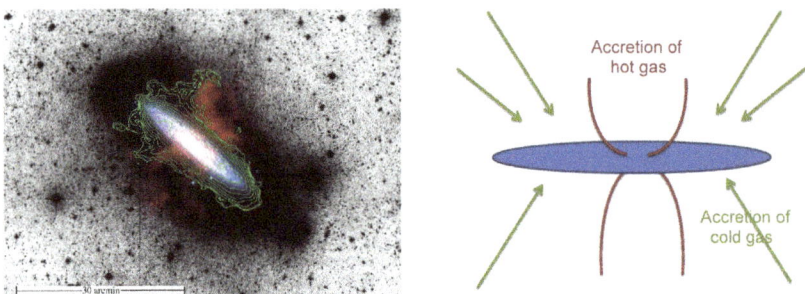

FIG. 8.9 – Example of simultaneous observation of hot gas ejection, from the intense star formation in the galaxy NGC 253, and of accretion of cold gas from intergalactic space. The green contours are those of the atomic gas emission at 21 cm. The diffuse emission in red comes from X-rays emitted by the ejected hot gas at a million degrees. The whole is superimposed on a deep optical image: black color for low brightness regions, and blue for the main optical disk. From Boomsma, R. et al. (2005), *Astronomy & Astrophysics* 431, 65-72, with permission of ESO. The drawing at the right schematically shows what happens.

# APPENDIX

## List of the principal Milky Way satellites, sorted by increasing distance

The existence of the Canis Major dwarf is controversial. It could be only a concentration of galactic stars

| Name | Type | Distance (kpc) | Absolute Magnitude | Apparent Magnitude |
|---|---|---|---|---|
| Canis Major Dwarf ? | Irr ? | 7.5? | ? | ? |
| Sagittarius Dwarf | dSph/E7 | 24 | −12.67 | 5 |
| Ursa Major II Dwarf | dSph | 30 | −4.2 | 14.3 |
| Large Magellanic Cloud (LMC) | Irr/SB(s)m | 50 | −17.93 | 0.9 |
| Boötes Dwarf | dSph | 60 | −5.8 | 13.1 |
| Small Magellanic Cloud (SMC, NGC 292) | SB(s)m pec | 63 | −16.35 | 2.7 |
| Ursa Minor Dwarf | dE4 | 63 | −7.13 | 11.9 |
| Draco Dwarf (DDO 208) | dE0 pec | 79 | −8.74 | 10.9 |
| Sextans Dwarf | dSph | 86 | −7.98 | 12 |
| Sculptor Dwarf (E351-G30) | dE3 | 88 | −9.77 | 10.1 |
| Ursa Major I Dwarf (UMa I dSph) | dSph | 100 | −6.75 | 13.5 |
| Carina Dwarf (E206-G220) | dE3 | 100 | −8.97 | 11.3 |
| Fornax Dwarf (E356-G04) | dSPh/E2 | 140 | −11.5 | 9.28 |
| Leo II Dwarf (Leo B, DDO 93) | dE0 pec | 215 | −9.23 | 12.45 |
| Leo I Dwarf (DDO 74) | dE3 | 250 | −10.97 | 11.18 |
| Leo T Dwarf | G | 420 | −7.12 | 16 |
| Phoenix Dwarf (P 6830) | IAm | 440 | −10.22 | 13.07 |
| NGC 6822 (Barnard's Galaxy) | IB(s)m IV-V | 500 | −15.22 | 9.32 |
| Cetus Dwarf | dSph/E4 | 750 | −10.18 | 14.4 |
| Leo A (Leo III, DDO 69) | IBm V | 800 | −11.68 | 12.92 |
| Tucana Dwarf | dE5 | 880 | −9.16 | 15.7 |
| UGC 4879 (VV124) | IAm | 1250 | −11.5 | 14.0 |
| Antlia Dwarf | dE3.5 | 1250 | −9.63 | 16.19 |

# Chapter 9
# The future

The considerable progress that has been made over the last twenty years in the knowledge of the Milky Way is based in part on the use of results obtained with the satellite HIPPARCOS and large surveys from the ground, particularly in the infrared, and secondly on high-performance simulations permitted by modern computers. However, numerous questions remain, as we have seen in previous chapters. Answers should be given to most of them with the new means of observation that are becoming available to astronomers, or with those that will be available in the near future.

We have learned a lot about the spiral structure of the Galaxy, its warping in the outer parts, the flaring of the plane, and gas accretion by high velocity clouds through mapping of atomic hydrogen by its emission line at 21 cm. In the future, new technology radio telescopes will be faster, more sensitive, and have higher spatial resolution. The international project SKA (Square Kilometer Array) is the flagship project of this new generation of radio interferometers (Fig. 9.1). With a sensitivity close to 100 times greater than current telescopes (total area of one square kilometer, or one million square meters), at frequencies between 0.15 and 25 GHz (wavelengths of 1.2 cm to 2 m), it will be characterized by a very large field of view (1 to 100 square degrees according to the wavelength), and a high spatial resolution (up to 10 milliseconds of arc). It is scheduled to start operations in 2020 and its elements will span two continents, which are relatively immune to radio interferences related to telecommunications, including mobile phones: South Africa for short wavelengths and Australia for the longer ones. This radio telescope of the future uses phased arrays, which can synthesize the observation beam through electronically: the actual antennas can be fixed and cheap, and the observation is based on a very powerful electronics that can form several beams to observe in different directions simultaneously. Thus eight different fields of view can be observed at the same time with SKA.

Star formation in molecular clouds remains somewhat mysterious because it is difficult to observe directly. This is especially true for massive stars, which are buried in a very opaque cocoon that dissipates only when they have reached the main sequence. Already, infrared satellites and mainly the HERSCHEL satellite (Fig. 9.2), which observed where interstellar matter becomes transparent in submillimeter waves, brought us surprising information about the formation of stars. But these satellites lack angular resolution. This resolution, absolutely necessary to see the details of what happens during the formation of a star, will be improved by the successor to the Hubble Space Telescope, the James Webb Space Telescope (JWST) to be launched in 2018, and especially by the interferometers in millimeter and submillimeter waves, whose sensitivity is also very significant: the greatest is in the southern hemisphere, ALMA, the world's largest astronomy project on the ground (Fig. 9.3), and, in the northern hemisphere, the interferometer of the Institute of Millimeter Radio Astronomy (IRAM) Franco-German-Spanish (Fig. 9.4), which is currently being extended into NOEMA (NOrthern Extended Millimeter Array), whose total area will be one third of that of ALMA.

FIG. 9.1 – One element of the short wavelength part of the giant interferometer SKA (Square Kilometer Array). Ensembles like this one will be scattered all over South Africa, to obtain a very high resolving power.

FIG. 9.2 – The satellite HERSCHEL of the European Space Agency (ESA). The primary mirror was 3.5 m in diameter. Courtesy ESA.

*The future*

FIG. 9.3 – Some of the antennas 12 m in diameter of the large millimeter and submillimeter interferometer ALMA (Atacama Large Millimeter Array) built by the European Southern Observatory, the United States, Canada, Japan and Taiwan. This instrument located at 5000 m altitude in northern Chile, is now fully operational, and includes 54 antennas of 12 m diameter and 12 of 7 m diameter. Courtesy ESO.

FIG. 9.4 – The millimeter IRAM interferometer on the Plateau de Bure (Hautes-Alpes), at 2550 m above sea level, with 6 of its 7 antennae 15 m in diameter moving on two tracks. The construction of 5 new antennae is scheduled, to get to 12 antennae in total, leading to the NOEMA instrument. © IRAM/Rebus.

The HIPPARCOS satellite had its limitations: it has significantly improved the distance scale in the Galaxy and in the extragalactic universe, through direct measurements of star distances by a geometric parallax, as well as those of stellar proper motions. But those were limited to stars brighter than magnitude 12 and closer to a hundred parsecs, that is to say to the solar neighborhood: the thick disk and the bulge, for example, remained inaccessible. In addition, the satellite did not have an instrument capable of measuring the radial velocity of stars, and thus could measure only two components of their velocity in space: it is necessary to complete its data by spectroscopic measurements on the ground, which are necessarily long and tedious.

The GAIA satellite of the European Space Agency (Fig. 9.5) is presently taking over. Much more sensitive than HIPPARCOS, it reaches a magnitude 20 and measures the positions of stars with a precision of a few millionths of an arc second instead of a millisecond for HIPPARCOS, at least for fairly bright stars. This will provide distance measurements and the proper motion of stars in the entire Milky Way. GAIA also features two other instruments: a spectrophotometer that will provide the spectral energy distribution in the visible and near infrared of all observed stars, from which their metallicity can be derived for example; the other instrument is a spectrometer that will give the radial velocity of stars down to magnitude 16. It is a complete package perfectly suited to the study of stellar populations across the Milky Way: spatial distribution, kinematics and chemical composition. The end of the mission is scheduled for 2019.

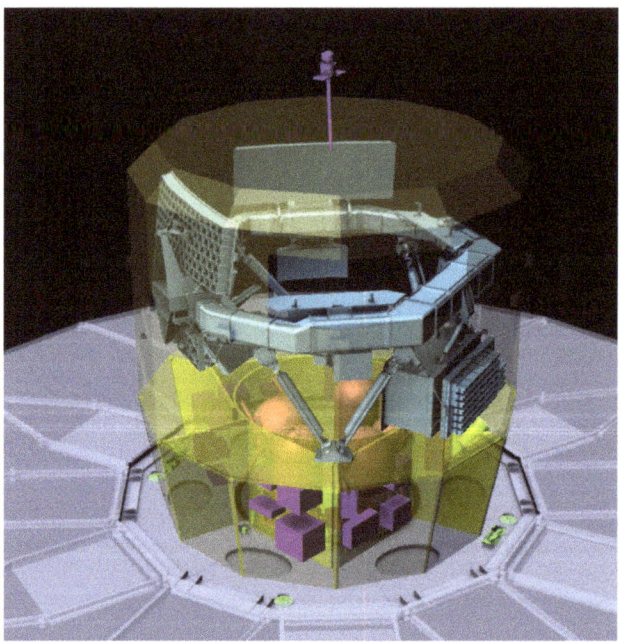

FIG. 9.5 – The GAIA satellite of the European Space Agency, has been launched in 2013. This transparent view through a protective envelope reveals in particular the optical bench almost octagonal (light blue), the essential part of the measuring system for star positions. One can see above the two rectangular mirrors (1.45 m × 0.5 m), which allow two fields of view at 106.5° to each other to be observed simultaneously. Courtesy ESA.

# The future

There will undoubtedly be the need to reach the kinematic parameters, and to achieve the spectroscopy of still fainter stars, or stars that interstellar extinction prevents us seeing in the visible. For this, the remarkable kinematic studies of the stars orbiting the central black hole of the Galaxy (see Chapter 4) show us the way: one can, from the ground, get proper motions relative to a bright object in the field of view with great precision thanks to the large telescopes equipped with adaptive optics. The giant telescopes of the future, like the E-ELT currently in construction by the European Southern Observatory (Fig. 9.6) will allow studies of this kind with a larger sensitivity and much greater accuracy and will also measure the radial velocity of objects inaccessible with GAIA. These are exciting prospects for the detailed knowledge of our Galaxy.

Fig. 9.6 – The telescope of 39 m in diameter E-ELT (European Extremely Large Telescope) under construction by the European Southern Observatory (ESO). Its first light is targeted for 2024. Courtesy ESO.

# Appendix 1
# Stellar parameters

TAB. A.1 – Parameters of stars of different types. The color $B$–$V$, the effective temperature $T_{\text{eff}}$, the absolute magnitude $M_V$ in the V filter and the luminosity $L$ in solar luminosities are given for the different types. WC and WN are the two main types of Wolf-Rayet stars.

| Type | Dwarfs (type V) | | | | Giants (Type III) | | | | Supergiants (Type I) | | | |
|---|---|---|---|---|---|---|---|---|---|---|---|---|
| | $B$–$V$ | $T_{\text{eff}}$ (K) | $M_V$ | $L$ (L$_\odot$) | $B$–$V$ | $T_{\text{eff}}$ (K) | $M_V$ | $L$ (L$_\odot$) | $B$–$V$ | $T_{\text{eff}}$ (K) | $M_V$ | $L$ (L$_\odot$) |
| WC | | | | | | | | | - | ≈ 50 000 | -8.5 | ≈ 2 × 10$^5$ |
| WN | | | | | | | | | - | ≈ 40 000 | -8.0 | ≈ 10$^5$ |
| O5 | -0.33 | 44 500 | -5.7 | 2 × 10$^5$ | | | | | -0.30 | 40 300 | -6.6 | 1.1 × 10$^6$ |
| B0 | -0.30 | 30 000 | -4.0 | 5.2 × 10$^4$ | | | | | -0.25 | 26 000 | -6.4 | 2.6 × 10$^5$ |
| B5 | -0.17 | 15 200 | -1.2 | 8.3 × 10$^2$ | | | | | -0.10 | 13 600 | -6.2 | 5.2 × 10$^4$ |
| A0 | -0.02 | 9 520 | 0.6 | 54 | -0.03 | 10 100 | 0.0 | 106 | -0.01 | 9 730 | -6.3 | 3.5 × 10$^4$ |
| A5 | 0.15 | 8 200 | 1.9 | 14 | 0.15 | 8 100 | 0.7 | 43 | 0.09 | 8 510 | -6.6 | 3.5 × 10$^4$ |
| F0 | 0.30 | 7 200 | 2.7 | 6.5 | 0.30 | 7 150 | 1.5 | 20 | 0.17 | 7 700 | -6.6 | 3.2 × 10$^4$ |
| F5 | 0.44 | 6 440 | 3.5 | 3.2 | 0.43 | 6 470 | 1.6 | 17 | 0.32 | 6 900 | -6.6 | 3.2 × 10$^4$ |
| G0 | 0.58 | 6 030 | 4.4 | 1.5 | 0.65 | 5 850 | 1.0 | 34 | 0.76 | 5 550 | -6.4 | 3.0 × 10$^4$ |
| G5 | 0.68 | 5 770 | 5.1 | 0.79 | 0.86 | 5 150 | 0.9 | 43 | 1.02 | 4 850 | -6.2 | 2.9 × 10$^4$ |
| K0 | 0.81 | 5 250 | 5.9 | 0.42 | 1.00 | 4 750 | 0.7 | 60 | 1.25 | 4 420 | -6.0 | 2.9 × 10$^4$ |
| K5 | 1.15 | 4 350 | 7.4 | 0.15 | 1.50 | 3 950 | -0.2 | 170 | 1.60 | 3 850 | -5.8 | 3.8 × 10$^4$ |
| M0 | 1.40 | 3 850 | 8.8 | 7.7 × 10$^{-2}$ | 1.56 | 3 800 | -0.4 | 330 | 1.67 | 3 650 | -5.6 | 1 × 10$^4$ |
| M5 | 1.64 | 3 240 | 12.3 | 1.1 × 10$^{-2}$ | 1.63 | 3 330 | -0.3 | 930 | 1.80 | 2 800 | -5.6 | 3.0 × 10$^5$ |

TAB. A.2 – Parameters of stars with solar chemical composition on the zero-age main sequence, and lifetime on the main sequence, from André Maeder. Isabelle Baraffe and Gilles Chabrier).

| Mass ($M_\odot$) | Luminosity ($L_\odot$) | $T_{eff}$ (K) | Lifetime (years) |
|---|---|---|---|
| 120 | $1.8 \times 10^6$ | 53 300 | $2.5 \times 10^6$ |
| 60 | $5.3 \times 10^5$ | 48 200 | $3.4 \times 10^6$ |
| 25 | $7.9 \times 10^4$ | 37 900 | $6.4 \times 10^6$ |
| 15 | $2.0 \times 10^4$ | 31 000 | $1.2 \times 10^7$ |
| 9 | $1 \times 10^3$ | 24 200 | $2.6 \times 10^7$ |
| 5 | 550 | 17 200 | $9.4 \times 10^7$ |
| 3 | 81 | 12 200 | $3.5 \times 10^8$ |
| 2 | 16 | 9 100 | $1.1 \times 10^9$ |
| 1.5 | 7 | 7 100 | $2.7 \times 10^9$ |
| 1 | 0.69 | 5 600 | $9.8 \times 10^9$ |
| 0.9 | 0.49 | 5 300 | $1.5 \times 10^{10}$ |
| 0.7 | 0.086 | 4 300 | $> 1.5 \times 10^{10}$ |
| 0.4 | 0.023 | 3 500 | $\gg 1.5 \times 10^{10}$ |
| 0.2 | 0.003 | 3 300 | - |
| 0.1 | 0.001 | 2 800 | - |

# Appendix 2
# A few basic notions concerning the observations of the interstellar medium

Many of these observations are concerned with the spectral lines of atoms and molecules contained in the interstellar medium. The basic parameter of a line is its absorption coefficient. This absorption coefficient $\kappa_\nu$ per unit volume and per unit frequency $\nu$ of a spectral line is given by the general expression:

$$\kappa_\nu = c^2 n_l g_u / 8\pi \nu_0^2 g_l \, A_{ul}[1 - g_l n_u / g_u n_l] \, \Phi_{ul}(\nu) , \qquad (A2.1)$$

where c is the velocity of light, $\nu_0$ the central frequency of the line, $n_l$ and $n_u$ the number of atoms per unit volume, respectively in the lower and the upper level of the transition, $g_l$ and $g_u$ the respective statistical weights of these levels and $A_{ul}$ the probability of spontaneous emission of the transition. $\Phi_{ul}(\nu)$ is the normalized spectral distribution of the intensity of the emission line: $\int \Phi_{ul}(\nu) d\nu = 1$. The term $g_l n_u / g_u n_l$ is the correction for stimulated emission. It is generally negligible for optical lines because the upper level is only poorly populated (except in the case of lasers for which there is no astrophysical equivalent), but is fundamental for a radio line for which the populations of the two levels are comparable.

It is most often necessary to consider the *column density* $N$ of the considered atoms, which is the number of atoms in a column per unit section:

$$N = \int n \, ds, \text{ where } ds \text{ is the length element.} \qquad (A2.2)$$

We define the optical depth $\tau$ as:

$$\tau_\nu = \int \kappa_\nu ds. \qquad (A2.3)$$

If we are in thermodynamic equilibrium, i.e. the levels of transition are in collisional equilibrium with the medium at the absolute temperature $T$, we can write:

$$n_u / n_l = g_u / g_l \, \exp(-h\nu_0 / kT) , \qquad (A2.4)$$

where h and k are respectively the Planck's and the Boltzmann's constant. At optical frequencies, $h\nu_0/kT$ is very small, but this is the reverse at radio frequencies for which one may write (the Rayleigh-Jeans approximation):

$$n_u/n_l \approx g_u/g_l \,(1 - h\nu_0/kT). \tag{A2.5}$$

This formula is valid for the 21-cm line of atomic hydrogen, which is actually at thermodynamic equilibrium. But it is not valid in the general case, so we define the population of levels using an excitation temperature $T_{ex}$ that has no physical meaning (it can even be negative if there is population inversion as is the case in the laboratory or interstellar masers):

$$n_u/n_l \approx g_u/g_l \,(1 - h\nu_0/kT_{ex}). \tag{A2.6}$$

To complete this basic summary, it is useful to introduce the notations used in radio astronomy. At decimeter and centimeter wavelengths, we are almost always in the Rayleigh-Jeans approximation so that the expression of the black body brightness is reduced to:

$$B_\nu(T) \approx 2kT\nu^2/c^2, \tag{A2.7}$$

Radio astronomers are accustomed to express the surface brightness' or intensities in terms of brightness temperatures $T_B$ defined by the above equation. Consequently, the brightness temperature in the 21-cm line of an atomic cloud with optical thickness $\tau$ and kinetic temperature $T_k$ is simply:

$$T_B = (1 - e^{-\tau})(T_k - T_{CMB}) \approx \tau(T_k - T_{CMB}), \tag{A2.8}$$

if $\tau$ is small. Here, we take into account the cosmic microwave background (CMB) radiation, which is that of a blackbody at $T_{CMB} = 2.726$ K: if the kinetic temperature is equal to $T_{CMB}$, the line is not detectable above the continuum background.

The Rayleigh-Jeans approximation is less and less valid when we go to the millimeter and submillimeter waves, but radio astronomers retain the habit of expressing the intensities in the form of fictitious brightness temperatures $T_B^*$, which can lead to confusion if we are not warned about it. The relationship between $T_B^*$ and the monochromatic brightness or intensity $I_\nu$ (above the cosmic continuum background) is then:

$$T_B^* = c^2/2k\nu^2 \, I_\nu. \tag{A2.9}$$

Furthermore, the expression (A2.8) is valid only when the medium is in thermodynamic equilibrium, which is, as we have said, the case of interstellar atomic hydrogen. This is a rather exceptional case, because in general the gas is far from thermodynamic equilibrium and density intervenes. We cannot dwell here on these complications.

# Glossary

*Accretion:* gravitational capture of matter by a star.

- *Accretion disk:* the matter accreted by a moving object is generally collected in a rotating flat disk, from which it falls slowly on the object.

*Apex:* the direction towards which the Sun moves relative to nearby stars.

*Astronomical unit:* a unit of length equal to the semi-major axis of the Earth's orbit (150 million kilometers).

*Black hole:* a very dense object whose gravitational field is such that no radiation can get out and manifests itself by its gravitational field or by the radiation from the matter it captures. These objects are predicted by the theory of general relativity; some are stellar, others much more massive, are found in the nuclei of galaxies. The central black hole in the Milky Way has a mass of 4.3 million solar masses.

*Brown dwarf:* a star with a mass lower than 0.08 times the mass of the Sun, where thermonuclear reactions cannot occur. Without production of energy, brown dwarfs are cold objects and emit very little light.

*Bulge:* a concentration of stars in the central region of a galaxy, which has the form of a spheroid, with a light distribution as $\exp(r^{1/4})$, and without rotation. If this concentration is flatter, with rotation and has an exponential light distribution, it is called a *pseudo-bulge*.

*Bubble (interstellar):* a large gas shell in expansion, resulting from the winds emitted by the stars of a central open cluster and by the explosion as supernovae of the most massive stars in this cluster.

*Cepheid:* a periodic pulsating variable star, massive, bright and evolved. There is a relationship between the period and the absolute luminosity of cepheids, which by comparison with their apparent brightness allows estimation of their distance (and also that of galaxies close enough to be able to observe their cepheids).

*Cloud:*

- *interstellar cloud:* the interstellar material is mainly distributed in "clouds" of greatly varying shape (often strips or filaments), with sizes ranging from a few parsecs to hundreds of parsecs and masses from a few solar masses to millions of solar masses;

- *molecular cloud:* a massive and relatively dense interstellar cloud, almost entirely composed of molecules; such clouds are the sites of star formation;

- *dark cloud:* synonymous with molecular cloud: indeed molecular clouds contain dust that absorb light of stars located behind and are observed as dark clouds, or areas of absence of stars in projection.

- *high-velocity cloud:* a cloud that is rapidly falling into the galactic disk, coming from the intergalactic medium or the halo of the Milky Way.

For pictures, see Chapter 1.

*Cluster:* a group of stars born at approximately the same time; there are *open, or galactic, clusters* close the plane of the Galaxy and generally young (less than a few billion years) and *globular clusters*, more compact and not concentrated in the galactic plane, which are mostly very old in the Milky Way (over 10 billion years). For pictures, see Chapter 1.

*Cosmic rays:* high-energy charged particles that fill the Galaxy; these are electrons, positrons and nuclei of all kinds. The cosmic rays that are received on the surface of the Earth are actually the result of the interaction of these particles with the Earth's atmosphere.

*Cosmology:* the study of the structure and evolution of the Universe

*Color-magnitude diagram:* see Hertzprung-Russell diagram.

*Dark matter:* a hypothetical material of unknown nature that would be observed only by its gravitational effects.

*Density wave:* a compression wave of matter, acting on the stars and interstellar matter of a galaxy, which is the cause of the spiral structure.

*Differential rotation:* the non-uniformity of the angular velocity of rotation in a galaxy, a star, a planet, etc.

*Dynamical friction:* gravitational braking of a massive body passing through an ensemble of smaller free masses, such as stars in a galaxy. It is this phenomenon that allows the merging between galaxies.

# Glossary

*Epicyclic frequency:* the frequency of the radial oscillation of a star around the position of equilibrium between gravity and the centrifugal force of the rotation around the galactic center.

*Fountain (galactic):* the mechanism by which the gas ejected into the halo of the Milky Way galaxy by stellar winds and supernovae cools and falls back on the disk.

*Galaxy:* an ensemble of stars, interstellar matter and dark matter of large mass and dimensions, relatively isolated from each other.

- *Barred galaxy:* a spiral galaxy with a bar in its internal regions;
- *Elliptical or spheroidal galaxy:* a galaxy with a spherical or elliptical shape with little apparent structure, relatively devoid of gas and young stars;
- *Irregular galaxy:* a relatively shapeless galaxy, very rich in gas and young stars;
- *Lenticular galaxy:* a flat galaxy without spiral structure;
- *Spiral galaxy:* a flat galaxy showing a spiral structure.

  The mass of galaxies is from $10^7$ solar masses (or less) to $10^{12}$ solar masses.

*General relativity:* the theory of gravitation due to Einstein, based on several assumptions, including the identity of the inertial mass and gravitational mass. The effects predicted by this theory, for example the deflection of light by a mass that produces gravitational lensing and microlensing, have so far been fully verified.

*Gould belt:* a ring of interstellar matter and young stars inclined to the galactic plane, with dimensions of a few hundred parsecs.

*Gravitational waves:* emitted by a massive body in motion, they are deformations of space propagating at the speed of light. Their existence is predicted by general relativity, and is verified in binary pulsars.

*Halo:* a more or less spherical ensemble surrounding a galaxy, including globular clusters, old stars and some gas.

*Heliosphere:* a cavity carved by the solar wind in the local interstellar medium.

*Hertzprung-Russell diagram:* (in summary HR diagram, or CM for color-magnitude diagram). A diagram in which the brightness of stars is plotted as a function of some tracer of their surface temperature. Initially this tracer was the spectral type of the star; then the color was used, for example the difference of magnitudes $B$ (blue, 450 nm wavelength) and $V$ ("visible" at 550 nm). The theory of stellar evolution allows us to build a HR diagram called "theoretical", in which the luminosity of the star integrated on all wavelengths (bolometric luminosity) is plotted as a function of its surface temperature.

*HII region:* synonymous with ionized gaseous nebulae.

*Hypernova:* a supernova resulting from the explosion of a very massive star in rapid rotation. The hypernovae explosions are causing some of the gamma-ray bursts. Synonym: collapsar.

*Interferometer:* an instrument made of several interconnected optical or radio telescopes, to obtain high angular resolution observations and maps of celestial objects.

*Interstellar dust:* small-sized grains (0.1 micrometer or less) consisting of silicates, graphite and ice, which are mixed with the interstellar gas and which make up about 1% of its mass. They absorb, scatter and polarize the light of stars and are the sites of formation by catalysis of certain interstellar molecules.

*Interstellar extinction:* weakening of the light of stars or of other distant objects under the effect of interstellar dust.

*Interstellar matter:* a medium composed of gas (ions, atoms or molecules), mixed with dust, located between the stars. It is a very heterogeneous environment, with "clouds" immersed in a dilute gas. The average density of the interstellar matter close to the Sun is about 0.5 atoms per cubic centimeter.

*Interstellar molecules:* we know today more than 150 molecules in the interstellar medium, some with up to 13 atoms. Some are present everywhere, but there are molecular clouds almost entirely composed of molecules, where the most complex ones are found.

*Interstellar reddening:* the change of the color of the stars due to selective extinction by the interposed interstellar dust.

*Light year:* a unit of length used in astronomy, which corresponds to the distance traveled in one year by light in a vacuum. It is $9.7 \times 10^{15}$ m, or 0.307 parsec (pc).

*Lindblad resonance:* the resonance that occurs when the epicyclic frequency of stars is twice the difference between the frequency of rotation around the galactic center and the rotation frequency of the density wave.

*Local standard of rest* (LSR): a fictitious point at the position of the Sun that would have a pure rotation around the galactic center.

*Magellanic Clouds:* two irregular galaxies, satellites of our Galaxy, at distances of 50,000 and 63,000 parsecs respectively for the Large and the Small Cloud. These are the closest galaxies to us, and their study can be conducted with a wealth of detail impossible to achieve in more distant galaxies.

*Magnitude:* the logarithmic scale to measure the brightness of a star;

- *Apparent magnitude:* measures the apparent brightness: $m = -2.5 \log F$ + constant, if $F$ is the luminous flux;

- *Absolute magnitude:* measures the intrinsic brightness; by convention, the absolute magnitude $M$ and the apparent magnitude $m$ of a star would be the same if it were at a distance of 10 parsecs (32.6 light years): $m - M = 5 - 5 \log D$, where $D$ is the object distance in parsecs.

*Main Sequence:* a locus of points in the color-luminosity diagram where the stars stay when they burn hydrogen in their central regions. The position of a star on the sequence and the duration of its stay depend on its mass as a first approximation, the most massive stars being the bluest, brightest and those that evolve faster.

*Metallicity:* the ratio of the mass of elements heavier than hydrogen and helium to that of hydrogen in a star, in the interstellar matter or in a galaxy.

*Microlensing:* a star or planet passing in front of an object, such as a star or a galaxy nucleus, deflects the light rays and produces a temporary amplification of its light. This is a general relativity effect.

*Modified Newtonian Dynamics (MOND):* a theory in which Newton's law of universal attraction is modified at very large distances. It offers an alternative to dark matter to explain the dynamics of galaxies, in particular their flat rotation curves.

*Nebula:* a very imprecise general term for a diffuse celestial object (for pictures, see Chapter 1);

- *dark nebula:* a mass of interstellar gas and dust opaque to light;

- *galactic nebula (or HII region):* mass of interstellar gas ionized by ultraviolet radiation of hot, young stars; ex. : The Orion Nebula;

- *gaseous nebula:* synonymous with galactic nebula and HII region;

- *reflection nebula:* interstellar matter made visible by the diffusion of light of a bright star, located inside or nearby, by the dust it contains;

- *planetary nebula:* a mass of gas ejected by a low-mass star at the end of its evolution and ionized by the radiation of the residual core of this star;
- *protosolar nebula:* a mass of gas and interstellar dust from which the Solar system formed.
- *protostellar nebula:* same, for a star.

*Nova:* a star increasing suddenly of brightness and decreasing gradually over a few weeks. Novae are very close double stars in which one component is a white dwarf: during its evolution, the other star ejects material that falls onto the white dwarf, warming considerably so that explosive thermonuclear reactions occur. Some novae are recurrent.

*Nucleosynthesis:* the formation of chemical elements by nuclear reactions in stars.

*Parallax:* an astronomical term often used to designate the distance of an object, usually expressed in parsecs;

- *geometric parallax:* obtained by triangulation using as a basis a large distance on the Earth or its orbit around the Sun;
- *photometric parallax:* obtained by comparing the apparent magnitude of a star with its absolute magnitude determined from its spectral characteristics;
- *statistical parallax:* obtained by using the global kinematic properties of a group of stars moving together.

*Parsec:* a unit of length widely used by astronomers, such that the semi-major axis of Earth's orbit is seen at the distance of 1 parsec over an angle of one arc second. 1 parsec = 3.26 light year = $3.08 \times 10^{16}$ m.

*Photodissociation region:* a region below the surface of a neutral cloud subjected to ultraviolet radiation, such that only the elements of lower ionization potential than hydrogen are ionized, while most of the molecules are photodissociated.

*Precession:* the movement of the axis of a rotating body, which describes a cone under the influence of external forces; also, rotation of the orbit of a planet or a star.

*Proper motion:* the lateral movement of a star in the sky.

*Pulsar:* a stellar object emitting perfectly periodic radio pulses (and/or sometimes X-rays, optical or gamma-ray pulses). The period of pulsars is

# Glossary

from a few milliseconds to a few seconds; they are neutron stars in very fast rotation.

*Radial velocity:* the velocity of approach or recession of a star, counted positively in the case of recession.

*Radio astronomy:* the branch of astronomy that studies the radio emissions in the Universe. The Sun, the planets, some stars, the atomic, ionized or molecular interstellar gas, the high-energy cosmic ray electrons, the pulsars, galaxies and quasars emit radio waves.

*Radiogalaxy:* a galaxy, generally elliptical, which emits an intense radio emission by the synchrotron radiation mechanism.

*Radio source:* a cosmic source of radio waves, more or less extended.

*Rotation curve:* for a flattened galaxy, the law describing the variation of the rotational speed with radius.

*Spectral line:* the reinforcement or decrease in intensity in the spectrum of an object occurring at a specific wavelength; the line is in emission if there is reinforcement, and in absorption if there is a decrease. The wavelength of a line is characteristic of the atom, ion or molecule that produces it.

*Star:*

- *Neutron star:* a very dense star (the Sun's mass within 10 km, or one billion tons per $cm^3$) whose material is degenerate, being composed mostly of neutrons. Pulsars and some X-ray sources are neutron stars, a residue from the explosion of supernovae.

- *Double (or binary) star:* about half of the stars are in pairs. Close double stars, which are more or less in contact, are the site of very interesting phenomena that change their evolution vis-à-vis that of isolated stars. Novae, X-ray sources, etc., are such binary stars.

- *Giant:* a star in an advanced stage of evolution, which begins to "burn" helium and carbon, and whose envelope is extended and relatively cold. This is the stage that follows the station on the main sequence. The giants called asymptotic branch giants are in the latest stage of their evolution.

- *Dwarf:* a star of the main sequence, of relatively small mass and dimensions (e.g. the Sun). White dwarfs, however, are stars at the end of their evolution.

- *Supergiant:* a very massive and luminous star leaving or having left the main sequence.

- *Wolf-Rayet:* an emission line star that represents the final stage of the evolution of some very massive stars.

*Stellar population:* a set of stars characterized by a roughly similar chemical composition. Population II stars are very old and poor in heavy elements, the Population I stars of all ages are rich in heavy elements. This classification is simplistic and tends to be abandoned.

*Stellar wind:* mass loss by either very massive stars, where the wind is fast and due to the radiation pressure on the atoms of the envelope, or by red giants and supergiants, where the wind is slower and is due to the radiation pressure on the dust condensed in the atmosphere of the star.

*Supernova:* an explosion ending the lives of massive stars (more than 6 to 8 solar masses: SN Ib, Ic or SN II) or of certain close binaries (SN Ia). Several supernovae have been historically observed in the Galaxy. After the explosion, in some cases, a neutron star remains which may be a pulsar, or a black hole. The gas ejected at high speed forms a *supernova remnant* with particular optical, X-ray and radio properties (for pictures, see Chapter 1). Supernovae, or more precisely their remains, are the sources of cosmic rays.

*Tide:* a deformation produced on an object (planet, star, galaxy) by the gravitational action of a neighboring body.

*Turbulence:* agitation of a medium as random eddies. This is a situation frequently encountered in nature, particularly in the interstellar medium and in the outer layers of the stars.

*Very Long Baseline Interferometry (VLBI):* radio interferometry with very long baselines, often between radio telescopes located on different continents, which enables us to obtain a very high angular resolution and highly precise astrometry.

*White dwarf:* a very dense star (density of the order of one ton per $cm^3$), with a mass always less than 1.4 solar mass. This is the final stage in the evolution of low-mass stars (less than about 6 to 8 solar masses). The material is in the form of atomic nuclei and electrons, degenerate in the quantum sense. White dwarfs are faint objects due to their small size.

# Bibliography

Bertin G. (2000) *Dynamics of galaxies*, Cambridge University Press

Bertin G., Lin C.C. (1996) *Spiral Structure in Galaxies: A Density Wave Theory*, MIT, Library of Congress

Binney J., Merrifield M. (1998) *Galactic Astronomy*, Princeton University Press, Princeton

Binney L., Tremaine S. (2008) *Galactic Dynamics*, 2nd edition, Princeton University Press, Princeton

Bok B.J. (1981) *The Milky Way: Fifth Edition*, Harvard University Press

Clark S. (2008) *Galaxy: Exploring the Milky Way*, Fall River Press

Combes F., Boissé P., Mazure A., Blanchard A. (2002) *Galaxies and Cosmology*, Springer, 2nd edition

Combes F. (2010) *Mysteries of galaxy formation*, Springer

Gilmore G., King I.R., van der Kruit P.C. (1991) *The Milky Way As a Galaxy*, University Science Books, U.S.A.

Hartmann D., Burton W.B. (1997), *Atlas of Galactic Neutral Hydrogen*, Cambridge University Press

Lequeux J., with the collaboration of Falgarone E. and Ryter C. (2005) *The Interstellar Medium*, Springer, Berlin & Heidelberg

Lequeux J. (2013) *Birth, Evolution and Death of Stars*, EDP Sciences & World Scientific, Singapore

Marochnik L.S., Suchkov A.A., Shukurov A., Yastrzhembsky I. (1995) *The Milky Way Galaxy*, Gordon & Breach publishers

Matteucci F. (2001) *The Chemical Evolution of the Galaxy*, Kluwer Academic Publishers

Melia F. (2003) *Black Hole at the Center of Our Galaxy*, Princeton University Press

Mo H., van den Bosch F., White S. (2010) *Galaxy Formation and Evolution*, Cambridge University Press

Pagel B.E.J. (1997) Nucleosynthesis and chemical evolution of galaxies, Cambridge University Press

Sparke L.S., Gallagher J.S. III (2000) *Galaxies in the Universe: an introduction*, Cambridge University Press

van den Bergh S. (2000) The *Galaxies of the Local Group*, Cambridge University Press

Waller W.H. (2013) *The Milky Way, an insider's guide*, Princeton University Press

Whittet D.C.B. (2002) *Dust in the Galactic Environment*, Institute of Physics, Series in Astronomy & Astrophysics


**Review articles**

Barnes J., Hernquist L. (1992) Dynamics of interacting galaxies, *Annual Review of Astronomy & Astrophysics*, 30, 705-742

Beck R., Brandenburg A., Moss D., Shukurov A., Sokoloff D. (1996) Galactic magnetism: recent developments and perspectives, *Annual Review of Astronomy & Astrophysics*, 34, 155-206

Binney J.J. (1992) Warps, *Annual Review of Astronomy & Astrophysics*, 30, 51-74

Buta R., Combes F. (1996) Galactic Rings, *Fundamentals of Cosmic Physics*, 17, 95-281

Cayrel R. (1996) The first generations of stars, *Astronomy & Astrophysics Review*, 7, 217-242

Combes F. (1991) Distribution of CO in the Milky Way, *Annual Review of Astronomy & Astrophysics*, 29, 163-193

Condon J.J. (1992) Radio emission from normal galaxies, *Annual Review of Astronomy & Astrophysics*, 30, 575-611

Courteau S., Cappellari M., de Jong R.S. (2014) Galaxy masses, *Rev. Mod. Phys.* 86, 47

Cox D.P. (2005) The three-phase interstellar medium revisited, *Annual Review of Astronomy & Astrophysics*, 43, 337-385

Fich M., Tremaine S. (1991) The mass of the Galaxy, *Annual Review of Astronomy & Astrophysics*, 29, 409-445

Freeman K., Bland-Hawthorn J. (2002) The new Galaxy; signatures of its formation, *Annual Review of Astronomy & Astrophysics*, 40, 487-537

Frisch P.C., Redfield S., Slavin J. (2012) The interstellar medium surrounding the Sun, *Annual Review of Astronomy & Astrophysics*, 49, 237-279

Hodge P. (1989) Populations in local group galaxies, *Annual Review of Astronomy & Astrophysics*, 27, 139-159

Ivezić Ž., Beers T.C., Jurić M. (2012) Galactic stellar populations in the era of the Sloan digital sky survey and other large surveys, *Annual Review of Astronomy & Astrophysics*, 50, 251-304

Jog C.J., Combes F. (2009) Lopsided spiral galaxies, *Physics Reports*, 471(2), p. 75-111.

Kalberla P.M.W., Kerp J. (2009) The HI distribution of the Milky way, *Annual Review of Astronomy & Astrophysics*, 47, 27-61

Kennicutt R.C. Jr., Evans N.J. II (2012) Star formation in the Milky way and nearby galaxies, *Annual Review of Astronomy & Astrophysics*, 50, 531-608

Kormendy J., Kennicutt R.C. Jr. (2004) Secular evolution and the formation of pesudobulges in disk galaxies, *Annual Review of Astronomy & Astrophysics*, 42, 603-683

Madau P., Dickinson M. (2014) Cosmic Star Formation History, *Annual Review of Astronomy & Astrophysics*, 52, 415

McWilliam A. (1997) Abundance ratios and galactic chemical evolution, *Annual Review of Astronomy & Astrophysics*, 35, 503-556

Melia F., Falcke H. (2001) The supermassive black hole at the Galactic center, *Annual Review of Astronomy & Astrophysics*, 39, 309-352

Morris M., Serabyn E. (1996) The Galactic center environment, *Annual Review of Astronomy & Astrophysics*, 34, 645-701

Putma M.E., Peek J.E.G., Joung M.R. (2012) Gaseous galaxy halos, *Annual Review of Astronomy & Astrophysics*, 50, 491-529

Sanders R.H., McGaugh S.S. (2002) Modified Newtonian dynamics as an alternative to dark matter, *Annual Review of Astronomy & Astrophysics*, 40, 263-317

Sofue Y., Rubin V. (2001) Rotation curves of spiral galaxies, *Annual Review of Astronomy & Astrophysics*, 39, 137-174

Toomre, A. (1977) Theories of spiral structure, *Annual Review of Astronomy & Astrophysics*, 15, 437

van der Kruit P.C., Freeman K.C. (2011) Galactic disks, *Annual Review of Astronomy & Astrophysics*, 49, 301-371.

Veilleux S., Cecil G., Bland-Hawthorn J. (2005) Galactic winds, *Annual Review of Astronomy & Astrophysics*, 43, 769-826

Wakker B.P., van Woerden H. (1997) High-velocity clouds, *Annual Review of Astronomy & Astrophysics*, 35, 217-266

Wyse R.F.G., Gilmore G., Franx, M. (1997) Galactic bulges *Annual Review of Astronomy & Astrophysics*, 35, 637-675

# On Internet

*Assembling the Puzzle of the Milky Way*, colloque du Grand Bornand (2011), EJP Web of Conferences Vol. 19, http://www.epj-conferences.org/index.php?option=com_toc&url=/articles/epjconf/abs/2012/01/contents/contents.html

# Index

**A**

Abundance of elements, 112, 123-124, 129-137, 141
Absorption coefficient, 167
Accretion, 169
    by the Galaxy, 111, 113, 127, 128, 134, 135, 139-147, 154-156, 159
    by the central black hole, 91-94, 99
Apex, 2, 19, 27, 32, 169
Association (stellar), 12, 34, 64, 105
Astronomical unit (UA), 3, 169
Asymmetric current, 25
Atacama Large Millimeter-submilllimeter Array (ALMA), 160

**B**

Bar, 85-86, 103-113, 141, 142
Black hole, 90-101, 169
Brown dwarf, 17, 78, 169
Bubble (interstellar), 16, 63-64, 169
    local, 24, 35
    N 70 (in LMC), 16
Bulge, 44-48, 75, 76, 78, 81, 85-86, 101, 109, 116, 122, 123, 131-137, 140-143, 148, 149, 152, 161, 169

**C**

Cepheid, 21, 38, 39, 129, 169
Champagne effect, 62
Chemical evolution, 121-137
Classification of galaxies 148
    of the Galaxy, 147-149

Cloud (interstellar), 9, 15
    atomic: see Interstellar medium, atomic
    dark, 4, 14, 170
    B 68, 14
    high-velocity, 134, 147, 154-155, 170
    molecular, 21, 33, 34, 39, 49, 50, 54-56, 60, 66, 79-80, 89-93, 96, 99-100, 116, 170
COBE (satellite), 44, 57, 85, 87
Collapse (gravitational), 9, 66, 96, 126, 144
Color-magnitude diagram: see Hertzprung-Russell diagram
Column density, 51, 57, 167
Cosmic rays, 11, 66-68, 80, 81, 170

**D**

Density (local), 21-24
Density wave, 9-10
Dimensions of Milky way, 1-5, 37-41
Disk, 21-29, 44-45, 79, 111, 134-137, 139-142
    rotation, 39-40, 49
    thick disk 48-49, 139-142
    thickness, 25, 49, 53-54, 59
    thin disk, 50, 139-142
    warp, 41-42
Distance (stars), 18-19
Dust (interstellar), 4, 6, 7, 10, 11, 14, 55-59, 65
    temperature, 57-58
Dynamical friction, 95, 97-98, 116, 140, 141, 143, 150, 170
Dynamics (galactic), 25-29, 103-119
Dwarf (star), 20, 24, 25, 165-166

## E

European Extremely Large Telescope (E-ELT), 163

## F

Formation
    of the Galaxy, 139-145
    of stars, 50, 92, 104, 115, 116, 122, 127, 134-137, 154, 156
Fountain (galactic), 64, 134, 155, 171

## G

GAIA (satellite) III, IV, 12, 24, 33, 48, 83, 137, 145, 154, 162
Galactic center, 85-101
    Black hole, 90-101
    Central cluster, 94-96
    Flares in, 92-94
    Interstellar matter in, 90-92
    Sgr A*, 92-94, 97
Galaxies (individual)
    Hickson 87-A, 109
    M 31 (Andromeda), 1, 147
    M 81, 73
    M 83, 74
    Magellanic clouds, 1, 16, 47, 78, 147, 149-152, 154, 157, 171
    NGC 253, 156
    NGC 891, 71, 155, 156
    NGC 1073, 114
    NGC 1097, 108, 115
    NGC 3992, 114
    NGC 4565, 109
    NGC 4762, 49
    NGC 6744, 10
    NGC 6822, 157
    Sagittarius dwarf, 46, 47, 131, 144, 145, 150, 152-154, 157
Gamma rays, 10, 11, 56, 59, 64, 69, 81, 172, 174
Giant star, 19, 20, 25, 123, 129, 132, 133, 152, 165-166, 175, 176
Globular cluster, 4, 5, 6, 13, 31, 47, 96, 97, 131, 132, 136, 144, 153
    47 Tuc, 13
    M 54, 153
Gould belt, 24, 34, 35, 171

Gravitational wave, 171

## H

Halo, 171
    of dark matter, 42, 76, 79-81, 121, 122, 151, 153
    gaseous, 64, 68, 92, 93, 134, 155-156, 170, 171
    radio, 70, 71
    stellar, 30, 31, 44-47, 96, 123, 127, 131-132, 135-136, 139, 140-145, 153
Heliosphere, 32, 35, 171
Herschel (satellite), IV, 159, 160
Hertzprung-Russell diagram, 17-21, 171
High-velocity cloud, 134, 154, 170
High-velocity star, 30-31
Hipparcos (satellite), III, 12, 19, 20, 23, 25, 27, 29, 159, 161, 162
HII region, 7, 14, 32, 38, 39, 41, 61-63, 65, 73, 81, 103, 104,, 105, 129, 130, 172 Orion
    Orion nebula, 17, 61, 173
    Trifid nebula, 14
Hypernova, 172

## I

Institute of Radio Astronomy at Millimeter wavelengths (IRAM), 160, 161
Interaction (between galaxies), 116, 149-150, 152-155
Interferometer, 159, 160, 161, 172
Interstellar medium, 31-36, 51-63, 87-88, 167-168, 172
    absorption lines, 4, 6, 32-34, 41, 54, 55, 64, 66, 77, 167-168, 175
    abundances, 54-55, 60, 61, 79, 113, 124, 128-130
    atomic, 4, 15, 33, 35, 41, 51, 52
    dust: see Dust
    extinction, 172
    hot, 11, 32, 33, 63-64, 65, 77, 134, 156
    ionized, 7, 8, 11, 32, 33, 51, 60-63, 66, 91-93, 172
    local, 31-35

magnetic field, 11, 63, 66-70, 72, 80, 94,
mass of, 32
molecular, 6, 8, 9, 11, 21, 32, 33-35, 50, 54-60, 62, 66, 79, 86-93, 106, 108, 116, 167
physical properties, 32
radiation field, 64-66
reddening: see extinction
turbulence, 62, 176
warm, 32, 35, 51, 52, 54, 88
Isotope, 68, 123, 133, 138

## J

*JWST* (satellite), IV, 160

## L

Lifetime (stars), 19, 20, 25, 166
Local standard of rest (LSR), 26, 173
Luminosity (stars), 3, 17-22, 165, 166
Luminosity function (stars), 3
Luminosity function (galaxies), 147-148

## M

Magellanic stream, 47, 149-150, 154
Magnitude, 18-19
Main sequence (stars), 19-20, 24-26, 29, 123, 132, 159, 165-166, 173
Map of the Galaxy, 7, 8, 9, 11, 40, 46, 54, 58, 63, 71, 72, 88, 154
Maser (interstellar), 41, 168
Mass (stars), 19, 22, 166
Mass (of the Galaxy), 6, 32, 42-44, 75-83
Mass-luminosity relation, 22, 77
Merging of galaxies, 47, 141-144, 170
Metallicity: see abundance of elements
Microlensing (gravitational), 78-79, 173
Migration, 50, 112-113, 127, 129, 132, 134-135, 139-142
Modified gravity (MOND), 42, 43, 82-83, 173
Molecule: see Interstellar medium, molecular

## N

Nebula
    dark : see Cloud (molecular, dark)
    gaseous: see HII region
    planetary, 15, 174
    protosolar, 174
    protostellar, 174
    reflection, 14, 173
Neutron star, 20, 24, 121, 175, 176
Nucleosynthesis, 77, 123-127, 174

## O

Oort's constants, 27, 37, 38
Open cluster, 4, 12, 13, 16, 50, 129, 130, 169
    NGC 3603, 13
Orbit (stars or gas), 25, 30, 31, 46, 85, 88-89, 94-96, 98-101, 105-106, 112, 114, 116-119, 135, 140, 144-145, 163

## P

PAH, 11, 65
Parallax: see Distance
Parenago's discontinuity, 25
Parsec (pc), 3
Photodissociation region (PDR), 61-62, 174
*Planck* (satellite), 57-59, 63, 71, 77
Population (stellar), 44-50, 95, 123, 129, 131, 132, 137, 139, 145, 162, 176
Precession, 89, 106, 114, 155, 174
Proper motion, 1-3, 12, 18-19, 22, 23, 27, 30, 41, 48, 92, 94-95, 101, 145, 150, 153, 154, 161-163, 174
Pulsar: see also Neutron star, 69, 171, 174

## R

Radial velocity, 6-8, 12, 22, 27, 37-38, 40, 51, 61, 73, 85, 99, 145, 150, 154, 162, 163, 175
Radioastronomy, 7, 176-168
Radiogalaxy, 175

Relativity (general), 78, 169, 171
Reddening: see interstellar medium, extinction
Resonance, 86, 88, 89, 101, 104-106, 108-110, 112-115, 118-119, 139, 142, 172
Ring, 10, 24, 47, 56, 75, 81, 86-88, 91-93, 101, 106, 108, 113, 115, 116, 134, 149, 153
    Monoceros ring, 81, 153
Rotation (galactic): see also Resonance, 6-8, 10, 21, 25-27, 30, 31, 37-44, 45, 47, 48, 72, 73, 75-83, 88, 91-92, 104, 114, 119, 135-136, 141, 142, 170

**S**

Satellites of the Galaxy, 47, 122, 139, 140, 142-145, 149-151, 154, 157
Square Kilometer Array (SKA), IV, 159, 160
Sloan Digital Sky Survey (SDSS), III, 47, 81, 151
Spectral line, 3, 19, 22, 51-56, 167-168
Spheroidal galaxies, 47, 131, 151, 171
Spiral structure, 6-9, 72-75, 103-108, 113, 115
Stellar current, 6, 25, 27-29, 46-47
Supergiant star, 21, 166, 176
Supernova, supernova remnant, 11, 16, 30, 35, 41, 50, 63-65, 68-71, 77, 78, 83, 92, 93, 104, 123-128, 134, 143, 155, 176
    Cassiopeia A, 16
Synchrotron radiation, 11, 66, 68-71, 92, 93

**T**

Temperature
    brightness (radio), 51, 52, 168
    effective (stars), 17-20, 66-67
    excitation, 168
Tide, 176
Tully-Fisher relation, 83, 147, 148
Turbulence, 62, 94, 176
Twenty-one centimeter line, 6-7, 11, 33, 37-40, 51-52, 56-58, 73, 80, 107, 134, 149, 154-156, 168
Two Micron All-Sky Survey (2MASS), 44, 85

**U**

*ULYSSES* (probe), 32

**V**

Velocity dispersion (stars, gas), 21, 23, 25-27, 29, 49, 73, 80, 86, 103, 110-112, 135, 141
Very Long Baseline Interferometry (VLBI), 41, 94, 176
Vertex deviation, 27, 28, 29

**W**

White dwarf, 20, 24, 63, 77, 78, 121, 124-126, 174, 176
WIMP, 81, 82
Wind
    galactic, 121, 154-155
    solar, 31-32, 67, 68, 70
    stellar, 16, 50, 64, 91, 92, 98, 104, 124, 125, 127, 143, 176
Wolf-Rayet star, 21, 95, 165, 176

**X**

X-rays, 10, 11, 33, 51, 64, 69, 77, 92, 99, 123, 156

www.ingramcontent.com/pod-product-compliance
Ingram Content Group UK Ltd.
Pitfield, Milton Keynes, MK11 3LW, UK
UKHW020239240426
12049UKWH00007B/131